THE HULETT HOTEL FIRE ON LAKE GEORGE

The Hulett House Hotel as it appeared in 1911. *U.S. Library of Congress.*

THE HULETT HOTEL FIRE ON LAKE GEORGE

GEORGE T. KAPUSINSKI

THE History PRESS

Published by The History Press
Charleston, SC 29403
www.historypress.net

First published 2012

Manufactured in the United States

ISBN 978.1.60949.261.8

Library of Congress Cataloging-in-Publication Data

Kapusinski, George T.
The Hulett Hotel fire on Lake George / George T. Kapusinski.
p. cm.
Includes bibliographical references.
ISBN 978-1-60949-261-8
1. Huletts Landing (N.Y.)--History--20th century. 2. George, Lake, Region (N.Y.)--History--20th century. 3. Hulett Hotel (Huletts Landing, N.Y.)--Fire, 1915. 4. Fires--New York (State)--Lake George--History--20th century. 5. Arson--New York (State)--Lake George--History--20th century. 6. Murder--New York (State)--Lake George--History--20th century. 7. Trials--New York (State)--Lake George--History--20th century. 8. Huletts Landing (N.Y.)--Biography. I. Title.
F129.H87K36 2012
974.7'49--dc23
2011046696

This book is dedicated to all those who have built, and all those who continue to build, the places we call home.

Contents

Preface 9
Introduction 11

1. Beginnings 13
2. Places of Interest 19
3. The Glenwood Hotel and Fire 29
4. Bought and Sold 35
5. The Huletts Landing Fire 41
6. The New Hulett Hotel 49
7. Suspicion 65
8. The Arrests 71
9. The Cast 77
10. The Trial 85
11. The Pictures 129
12. Aftermath 131

Epilogue 147
Bibliography 151
About the Author 157

Preface

Huletts Landing, New York, is a relatively quiet place. Ask anyone who lives or visits here and you will most likely get unanimous agreement that little happens in Huletts most of the year. It is primarily a summer destination. I think most would probably say that the reason they come to Huletts, besides the scenic beauty of Lake George, is for the peace and quiet.

That's not to say that things don't happen here. It's just that there are plenty of places where the mix of people and activity has a faster pace, and as such, things happen that make news at a greater frequency than in Huletts.

Looking back over the last one hundred years, I've always believed that the biggest event that happened in Huletts was the fire that burned the Hulett Hotel in 1915. The resulting news was a sensational arson trial that was picked up by many local papers during that time. The rebuilding of the Hulett Hotel in a slightly different location started a series of events that, in many ways, forms the identity of Huletts Landing today.

I always wished that there was a time capsule from 1915 that would answer the many questions I have about what happened on that horrendous day when the first Hulett Hotel burned. Likewise, I have searched for years—to no avail—for the blueprints of the second hotel that was built to take its place.

The year 1915 was a long time ago. The historical record doesn't leave much about what happened in a quiet place like Huletts in an era when news traveled much more slowly than it does today. The people who were alive then are not alive now, and their recollections are lost to antiquity. Likewise, photography was still in its infancy, and pictures from that time are hard to come by. Historical documents regarding this tiny hamlet, in my estimation, are even more rare.

After writing my first book on the history of Huletts Landing, which was really a compilation of photographs that detailed the history of Huletts, I truly believed that would be my first and last book. I felt I had exhausted the historical pictures in my family's collection, and I believed there was nothing more I could add to my narrative. It was an exhaustive work that took me many years to compile, and it was an exercise that, quite frankly, I never saw myself repeating.

That all changed, though, when some pictures came into my possession that literally took my breath away. This book is the story of those pictures, and the fact that they center on the rebuilding of the Hulett Hotel after the fire of 1915, I believe, makes this a very compelling story.

It is not my intent to prove the innocence or guilt of any particular person(s) or challenge the verdict issued in the arson trial that resulted from the fire. While I have my own opinion, which I will lay out in the succeeding pages, my goal here is to share history. What history is that, you might ask. Well, as you will read, some of the images presented here have their own unique stories that coincide with the fire and the rebuilding of the hotel during the winter of 1915–16. In order to tell these stories, the respective histories of the fire, arson trial and individuals involved have to accompany them.

Unlike with my first book, the words for this book jumped off my pen—or my computer, as the case may be—but there are some questions I couldn't answer and which remain mysteries. I do my best to present these lingering mysteries and will leave you to find their answers on your own. You may draw different conclusions from mine, but that's what makes history interesting.

The work that follows is part mystery, part history and part story of revelation. I have enjoyed writing it. I hope you will enjoy reading it.

Introduction

Lake George Resort Is Sold
Experienced Hotel Man Buys Picturesque Property

W.H. Wyatt of Troy has purchased from H.W. Buckell the property known as Huletts at Huletts Landing on Lake George. The transaction includes the hotel, fifteen cottages, a store near the boat landing, a saw mill and 120 acres of land. The consideration was $50,000. Huletts is one of the best known resorts on Lake George, and for many years has been one of the most popular of the numerous summer resorts on the lake. Mr. Buckell has owned and been in control of the property for twenty seven years. Mr. Wyatt will take possession April 1 and proposes to make many improvements to the hotel and cottages. He will give his personal attention to the management of the hotel.

The resort has been popularly called "Picturesque Huletts" because of the magnificent scenery, both on land and lake in the nearby vicinity. The hotel and cottages are located eighteen miles down the lake on the east shore and in the midst of one of the prettiest spots of Lake George. Nearby is the Elephant, a mountain rising to a height of 1,700 feet, and just beyond this is Black Mountain, 2,600 feet in height, making an impressive background for the beautiful view spread before the hotel.

Mr. Wyatt is an experienced hotel man, having been for a number of years the proprietor of the Clark house in Troy. He also owned and conducted the Glenwood at Lake Bomoseen, which was destroyed by fire

> *September 23, 1912. Under the direction of Mr. Wyatt, Huletts will maintain the high standard it has enjoyed among summer resorts and if it be possible it will be raised.*
>
> —Ticonderoga Sentinel, *January 30, 1913*

Note: While the above newspaper article says that the Glenwood Hotel on Lake Bomoseen was destroyed on September 23, 1912, the actual date of the fire was September 17, 1912.

1

Beginnings

The illustration is of a poor black sharecropper appealing to Abraham Lincoln, with the man's wife and children standing in the doorway of a nearby log cabin. The American flag hangs touching the ground on the left side, while Union and Confederate troops battle in the background for the soul of a young nation. Lincoln is pictured holding his top hat in his right hand. Perhaps this is an illustration of the same top hat he wore to Ford's Theatre on the fateful night of April 14, 1865. The light streams into the center of the illustration, perhaps signifying hope.

Underneath the depiction, the following words appear:

Published by Gwendolyn Publishing Co., New York

Appealing to be Allowed to Help Fight for the U.S.
-or-
The Condition in 1863.

Its dimensions are only approximately eighteen inches high by thirty inches wide. The artist's name, J.E. Taylor, and the date, 1892, are clearly visible in the lower right-hand corner. The Abraham Lincoln Presidential Library and Museum in Springfield, Illinois, doesn't even have a copy of the drawing. The library's curator presumes that the artist's initials refer to James E. Taylor, an illustrator who did similar sketches for *Frank Leslie's Illustrated Newspaper*.

The sketch was framed, hung and finally stored. It was almost thrown away. But that is where this book begins. It begins here because any really good story has a mystery attached to it. This is the mystery of this story.

It is not the front of this drawing that is important here; rather, it is what was glued to the back of the drawing that this book will focus on.

Attached to the back of this depiction of Abraham Lincoln—hidden, ignored and almost discarded—were nineteen pictures, perfectly preserved, dating back to the winter of 1915–16, showing the aftermath of the Huletts Landing hotel fire of 1915 and the building of the next Hulett Hotel. The events that the nineteen pictures document prefigure a sensational arson trial that took place in May 1917 in which the hotel's owner, William H. Wyatt, was charged for allegedly having the hotel burned down.

LOST AND FOUND

While the new Hulett Hotel was taken down in the winter of 1959–60, the dormitory that housed its employees stood for another forty years. Dilapidated and run-down, it remained a testament to a wonderful time when the sounds of music drifted from the lakeside Huletts Casino and the guests at the hotel frolicked on the shoreline of Lake George.

The Hulett Hotel employee dormitory in the mid-1980s. The Lincoln illustration was tossed from a window as the dormitory was being cleaned of debris. *Kapusinski family collection.*

When the Town of Dresden closed its local landfill under mandates from New York State to local municipalities in the mid-1980s, my family, who owned the dormitory, started the process of cleaning the hulking structure of years' worth of accumulated debris.

While old mattresses, rubbish and assorted junk were thrown from windows and hauled out doors, a mildewed illustration in a broken frame was tossed from a second-floor window. It found a resting place in the high grass below the far end of the ramshackle old structure, just far enough removed from the hands disposing of the garbage that day. It wasn't until the next day, when the commotion had stopped and quiet had returned, that the old dormitory's nearest neighbor, Mr. Alex Manuele, wandered over. That's when a gleam from the frame caught his eye.

"To be quite honest, I didn't even see the illustration of Lincoln when I first saw it," recounted Mr. Manuele. "It was resting facedown, and what caught my attention was the frame, which I thought was quite nice and which I thought might be repaired. I assumed that it wasn't wanted, and I picked it up to work on it."

It wasn't until many years later, when Mr. Manuele went about trying to repair the frame, that he took out the illustration of Lincoln and found the pictures from the winter of 1915–16 glued to the back.

"When I realized what they were, I wanted to return them to you," recounted Mr. Manuele in a conversation with me during the summer of 2010. "They belong to your family, and if there was someone who would tell their story, it would be you."

And that's where this book begins—with an illustration of Lincoln that was almost thrown away and then was found but stored again for many years, with pictures glued intentionally to the back.

1861

William Seward, president Lincoln's secretary of state and former governor of New York. Seward was severely injured in an assassination attempt the night Lincoln was murdered. *U.S. Library of Congress.*

In 1861, Abraham Lincoln was inaugurated as the sixteenth president of the United States. This was also the year he chose William Seward—the twelfth governor of New York, a U.S. senator and one of Lincoln's rivals for the 1860 Republican nomination—to serve as his secretary of state. Seward was known as a fierce abolitionist.

Seward developed his views about slavery while he was still a boy. His parents, like other residents of the Hudson Valley in the early nineteenth century, owned several slaves. (Slavery was slowly abolished in New York between 1797 and 1827 through a gradual, mandated process.) He discerned very quickly the inequality between the races, writing in later years, "I early came to the conclusion that something was wrong…and [that] determined me…to be an abolitionist." This belief would stay with Seward through his life and permeate his career.

Seward is remembered for surviving an assassination attempt the same night Abraham Lincoln was shot. On April 14, 1865, Lewis Powell, an associate of John Wilkes Booth, gained access to Seward's home by telling a servant, William Bell, that he was delivering medicine for Seward, who was recovering from a broken jaw he had received in a carriage accident. Powell had just started up the stairs when he was confronted by one of Seward's sons, Frederick. Frederick told the intruder that his father was asleep, and Powell turned back down the stairs. Then he suddenly swung around and pointed a gun at Frederick's head. After the gun jammed, Powell panicked but repeatedly struck Frederick in the head and face with the pistol, leaving his victim in critical condition on the floor.

Powell then burst into William Seward's bedroom with a knife and stabbed him several times in the face and neck. Seward survived because the wire mesh holding his broken jaw together prevented the knife from cutting arteries in his face and neck. Powell then attacked Seward's son Augustus; a soldier named Sergeant George Robinson, who had been assigned to stay with Seward; and a messenger, Emerick Hansell, who arrived just as Powell was escaping.

Fortunately, all the men who were injured that night survived, although the first doctor on the scene said he had never seen so much blood. William Seward would carry the facial scars from the attack through the remainder of his life. Powell was captured the next day and was executed on July 7, 1865, along with David Herold, George Atzerodt and Mary Surratt, three other conspirators in the Lincoln assassination. Justice was swift in 1865.

William Seward would thus be inextricably tied to the presidency of Abraham Lincoln. This son of New York would be forever linked to perhaps the greatest leader America has ever produced. All of these events would be in the future, though, as the year 1861 dawned.

The year 1861 was also when another son of New York, William H. Wyatt, was born. His story and the small Adirondack hamlet of Huletts Landing on Lake George would likewise be bound together with Lincoln—not in the same way that William Seward and Abraham Lincoln were tied together historically, but pictures of Wyatt and an illustration of Lincoln would literally be bound together in a mystery that survives to this day.

2
Places of Interest

Narratives about the life of a man usually begin with the place he was born, continue through the towns where he lived and worked and conclude with an account of his death. While history leaves us many records in regard to some people, for others, the trail is more difficult to uncover. The narrative of this book will follow the life of a successful businessman who owned a hotel on Lake George that burned down and was rebuilt. Certain events concerning the rebuilding of that hotel were captured by a photographer, and these photos ended up on the back of a Lincoln illustration. This narrative is, in many ways, incomplete; history has left us few clues due to the length of time that has transpired.

Instead of focusing merely on the individuals involved in a narrative, sometimes an examination of the places where those people lived, and which shaped their characters and destinies, serves as a better guide to the people about whom we hope to learn. Places can tell us much about a person. Did they live in the city, where anonymity is expected, or did they love the interaction with neighbors that is more common in a rural setting? Complex characters are often forged in environments that are uniquely different, so a review of the places where individuals lived can be instructive.

The central characters of this work are complex, interesting people. They are all exceedingly fascinating figures whose lives, even one hundred years later, make for an interesting story.

To begin, let us examine the places that contributed to their characters and helped shape their destinies.

TROY, NEW YORK

William H. Wyatt was born at the very southern end of Washington County, New York, in 1861. He stated that he was born in Cambridge, New York, yet cemetery records say that his place of birth was in the neighboring town of Easton, New York. Few if any records exist about his early childhood and upbringing.

In the book *History of Washington County, 1878*, a William Wyatt from the town of Easton, who may have been William H. Wyatt's father, is shown to have enlisted in the Union army's 77th Regiment, Company K, on September 6, 1861, and was discharged on December 29, 1864. Most Cambridge men were in the 123rd Infantry during the Civil War. The reference index for the *Washington County Post* (1799–1986), the local Cambridge newspaper, lists a William Wyatt, age fifty-one, who died in Easton on April 13, 1877. Could this have been Wyatt's father? It is a distinct possibility.

In any event, Cambridge and Easton both have as their southern boundary the northern border of Rensselaer County. Troy is the seat of Rensselaer County and is located approximately sixteen miles south of Easton, so it is easy to see how Wyatt would have migrated south to the city of Troy, where he would spend most of his life operating hotels.

In the nineteenth century, Troy, at the junction of the Hudson River, the Erie Canal and several important railroad lines, was an industrial powerhouse. Its once great wealth was produced from the steel industry, with the first American Bessemer converter erected on the Wyantskill, a stream with falls in a small valley at the south end of the city. The industry first used charcoal and iron ore mined from the Adirondacks. Much of the iron and

Troy, as viewed from across the Hudson River looking east, circa 1909. Troy was where William H. Wyatt operated the Trojan Hotel. *U.S. Library of Congress.*

The grave of Samuel Wilson, also known as "Uncle Sam," is today a popular Troy tourist destination. *Creative Commons By-SA-3.0/Matt H. Wade at Wikipedia.*

steel produced in Troy was used by the extensive federal arsenal across the Hudson at Watervliet, then called West Troy.

The initial emphasis on heavy industry later spawned a wide variety of highly engineered mechanical and scientific equipment. Troy was the home of the W. & L.E. Gurley Co., maker of precision instruments. Gurley's theodolites were used to survey much of the American West after the Civil War and were highly regarded until laser and digital technology eclipsed telescope and compass technology in the 1970s.

It was this atmosphere that produced Samuel Wilson, a prosperous Troy meatpacker, whose name is purportedly the source of the personification of the United States as "Uncle Sam." During the War of 1812, Wilson obtained a contract to supply beef to the army and shipped it in barrels. The barrels, being government property, were branded with the initials "U.S.," but the teamsters and soldiers joked that the initials referred to "Uncle Sam," who supplied the product. Over a century later, on September 15, 1961, Congress adopted the following resolution: "Resolved by the Senate and the House of Representatives that the Congress salutes Uncle Sam Wilson of Troy, New York, as the progenitor of America's National symbol of Uncle Sam."

An illustration of the final chase of Moby-Dick. *From* Moby-Dick *(New York: Charles Scribner's Sons, 1902).*

Troy, under the patronage of Stephen van Rensselaer, was the home of the first strictly scientific academic institution in the United States: Rensselaer Polytechnic Institute, founded in 1824. This institute trained the corps of students who later founded the Massachusetts Institute of Technology.

Likewise, Herman Melville, the great American novelist, best known for his novel *Moby-Dick,* resided from 1838 to 1847 at what is now known as the Herman Melville House. One of the central characters of *Moby-Dick* is Captain Ahab, the tyrannical captain of the *Pequod* who is driven by a monomaniacal desire to kill Moby-Dick, the whale that had maimed him on a previous whaling voyage. During the final chase, Ahab hurls his last harpoon while yelling his now-famous line: "To the last I grapple with thee; from hell's heart I stab at thee; for hate's sake I spit my last breath at thee."

Thus, as the nineteenth century rolled into the twentieth, Troy was a place enlivened by commerce, science and culture. The population was recorded at 69,651, making the city an epicenter of business and activity. It was in exactly this type of environment that an entrepreneur like Wyatt could find success.

Wyatt seems to have purchased his first Troy hotel in 1901. Had he worked in other hotels? Did he have experience in the hotel industry? The facts are unclear because contemporaneous records that still exist are few. Newspapers of the day, however, do refer to William H. Wyatt as a "Troy hotel man." As the proprietor of a hotel, Wyatt would have come in contact with all sorts of interesting figures who would have been in Troy for all types of purposes.

Thus, it was Troy, New York—the city that produced the real "Uncle Sam" and the literary Captain Ahab, as well as some of the very science and technology that turned America into an industrial giant—that fashioned William H. Wyatt, the Troy hotel man.

Round Lake, New York

John D. Sharpe also plays a central role in the story that follows. Some newspaper accounts refer to him as living in the village of Round Lake, New York, and operating a livery stable.

The village of Round Lake is located in Saratoga County. In 1975, the Round Lake Historic District, which encompasses the village, was added to the National Register of Historic Places.

The J.M. Finch livery stable in Whitehall, New York. Fred Haney (left) and Job Finch pose with horses. Livery stables were once common sights in towns across America. *Courtesy of Mae Finch and the Whitehall Historical Society.*

The village of Round Lake began in 1867 as a summer camp for Methodists. It was modeled after Oak Bluffs, a Massachusetts town on Martha's Vineyard. At first, visitors lived in tents, and visiting ministers rented space to hold their summer religious activities. Many of those who came to Round Lake took the train from New York City. Members of the Methodist Church would come each summer to pitch tents and have open-air meetings. As these meetings became more and more popular, returning visitors began replacing the tents with small wooden buildings generally known as gingerbread cottages.

By the late 1800s, more permanent structures, including a two-story trustees' office and a market, appeared. The Round Lake Fire Company was founded in 1886 as the M.B. Sherman Hose Company, with a constitution that required active members to have "good moral character" and live within a mile of the community. The first firehouse was situated on Troy Avenue near the intersection of Burlington Avenue.

This atmosphere produced activity in the summer months when people from different areas arrived in Round Lake to experience religious fervor and revival.

Before the automobile became part of American culture, this was just the type of place where a livery stable could thrive. Defined as "the boarding and care of horses for a fee or the hiring out of horses and carriages," a livery was a bustling business in the early 1900s. A person employed in such an enterprise would meet people from all destinations and walks of life.

A livery stable would have been full of horses especially bred for carriage or road use. Carriage horses were more stylish in appearance, while road horses were more of a working breed. Both shared good conformation and a strong constitution. Horses were broken in using a bodiless carriage frame called a break or brake.

Little is known about John D. Sharpe. Can it be construed that a livery stable operator would be a good judge of horses? Was Sharpe a man who could tell if a horse was fast or slow simply by looking at it? It can be surmised from his occupation that he knew horses better than most. What is known is that Sharpe was born in approximately 1884 and lived throughout his life in numerous New York towns. Records indicate that he was married and had two children, but sometime after the birth of his second child, he left his wife and moved into a boardinghouse in Round Lake maintained by a woman known as Mrs. Cornelia Gries.

How would a married man who had left his wife be viewed in a place like Round Lake, which was known for its religious fervor, in the early 1900s? Sharpe's actions, if recognized by the local citizenry of Round Lake, would have most likely fit the classic definition of moral turpitude: "conduct that is considered contrary to community standards of justice, honesty or good morals." Was Sharpe's background even known by the local populace? History, once again, is unclear.

Still, livery stables were necessary for transportation during this time. Sharpe, like Wyatt, would have come into contact with all sorts of interesting people as he ran his stable and went about his day-to-day business in Round Lake.

In an interesting and historically unrelated coincidence that demonstrates how the places where people live sometimes intersect, William H. Wyatt also happened to own a farmhouse in Round Lake that burned down.

Lake Bomoseen, Vermont

While it cannot be said for certain how William H. Wyatt ultimately came to own the Glenwood Hotel on Lake Bomoseen, there is one notable connection worth exploring.

The Glenwood Hotel on Lake Bomoseen in Vermont was purchased by William H. Wyatt in 1912. *Courtesy of Joseph Doran.*

The lake itself is located on the western side of Vermont, near the border of New York. It is not far from Washington County, New York, which hugs the Vermont border, where William Wyatt spent his early childhood.

There is evidence that the original Native American inhabitants lived along the shores of the lake for at least six thousand years, as evidenced by the discovery of a Neville spear point. In 1609, Samuel de Champlain called the lake "Bombazine," but it wasn't until the 1860s that it was officially named by Robert Morris Copeland. One little-known fact about Lake Bomoseen is that it is the largest lake entirely within Vermont's boundaries.

Starting in the 1880s, numerous hotels and tourist attractions were built along the lake. In 1896, a man named William C. Mound opened the Glenwood Hotel and operated it for a number of years. The area was also known for its slate quarries, which can still be seen today. A Troy businessman named Joseph Billings owned a slate mill on the western shore of Lake Bomoseen, located near the Glenwood Hotel.

Wyatt would purchase the Glenwood in 1912 after Mound died suddenly during the summer of 1911. According to the *Fair Haven Era*, Mound had a fatal heart attack on August 30, 1911, when he was awakened at 2:00 a.m. by a raucous group of patrons who were returning from a "straw ride and… making a noisy demonstration…Hurriedly dressing himself he started up the stairs to quiet the disturbance but had no sooner reached the top flight [of stairs] before he fell dead."

Did the Billings family inform Wyatt of a sudden business opportunity? Was Wyatt familiar with Lake Bomoseen from his early days growing up in Washington County, New York? Once again, history is unclear. What is known is that Wyatt purchased the Glenwood Hotel in 1912 and brought in John D. Sharpe to operate the Glenwood's livery stable.

Huletts Landing, New York

Huletts Landing is a hamlet in the town of Dresden in northern Washington County, New York. A lakeside community on the east shore of Lake George, it is located in the Adirondack Mountains. Lake George is affectionately called the "Queen of American Lakes" for its natural beauty and crystal-clear water. It drains north into Lake Champlain.

Up until approximately 1874, the area was known as Bosom Bay. The present name derives from the Hulett family, who settled the area. David Hulett was born in Killingly, Connecticut, sometime between 1758 and 1762 and served under General James Wadsworth during the American Revolution. For three years, the unit in which Hulett served fought mainly in upstate New York. Oral history asserts that David Hulett fought heroically at the Battle of Saratoga, where he refused to leave the field despite an injury to his neck. The Hulett family settled near Lake George in about 1804, possibly because of a land grant offered to veterans in lieu of wages.

Successive generations of the Hulett family lived by farming the area until Philander Hulett, David's grandson, began to develop the area in the 1870s by building a steamship landing and submitting an application for a post office. The post office application is the earliest document to use the Huletts Landing name.

The Lake George Steamboat Company, which delivered mail as well as passengers, probably required either a post office or a tourism facility in order to justify adding a new stop to its schedule. The application received approval, and Philander Hulett built the post office shortly afterward, housed in the same building as the community's store. Thus the name Huletts Landing came into existence.

Once the steamship began stopping daily, Philander Hulett started taking in summer tourists at his family farm. The Hulett House Hotel quickly

Tourists from Lake George arriving at Huletts Landing in 1900. *U.S. Library of Congress.*

expanded and developed into one of many popular tourist destinations on Lake George.

As the years passed, Philander Hulett sold his interest in the property. By 1905, a self-made Dresden businessman, Henry W. Buckell, owned the ever-expanding Hulett House Hotel.

One final notable feature about Huletts is that there is only one road into the hamlet. County Route 6—which splits into its suffixed routes 6A and 6B, which in turn head north and south, respectively, along Lake George once over the mountain—is the only way to reach the hamlet by land.

While the mountains that ring Lake George are known for their beauty, they make the terrain difficult to traverse. Huletts Landing is actually very close to the Vermont border, being located in Washington County, New York, only fourteen miles due west of Lake Bomoseen.

3

The Glenwood Hotel and Fire

The Glenwood Hotel was located on Lake Bomoseen's west shore and by all accounts was situated in a charming spot overlooking the lake. Most guests arrived by boat from the opposite shore. The hotel was also accessible by a local lane named Creek Road. It accommodated 150 guests when fully occupied and had a bowling alley and excursions to Lake George.

When originally constructed in 1896, the Glenwood was one of the largest hotels on the lake. It was 172 feet long and 72 feet wide, including a large veranda that overlooked almost 150 feet of waterfront. The *Fair Haven Era* reported in 1896 that

> *the Glenwood is a three story building containing about 40 large airy rooms all done in natural wood. The first floor is taken up with a large office, a parlor looking directly upon Bomoseen and the large and spacious dining room from the windows of which can be seen the lake. Water for drinking is to be supplied by a large spring while a large tank will constantly be kept filled with water by a six horse power engine which will furnish relief in case of fire.*

At the time William Wyatt purchased the Glenwood Hotel, the first floor of the four-story building contained the dining room, parlors, kitchen and dance hall. The upper three floors contained the bedrooms. Wyatt made numerous improvements during the spring of 1912, including painting all the buildings.

The Glenwood Hotel was a large structure with beautiful views of Lake Bomoseen. *Courtesy of Donald H. Thompson.*

The boat dock in front of the Glenwood Hotel brought visitors from the opposite shore. *Courtesy of Joseph Doran.*

In a 1912 brochure, which is considered quite rare today, Wyatt advertised that the Glenwood was "under new management and newly renovated." The brochure also emphasized that "the shade of the woods makes all the rooms in the house cool and comfortable even on the hottest days of mid-summer."

Here is how the brochure described the hotel:

> *The guests' room are all outside, and from every window a fine view either of the lake or the beautiful pines can be had. The verandas are broad and well shaded at all hours and are twelve feet wide with frontage of 172 feet and 72 feet on the sides. The entire interior is finished in natural wood, with polished floors.*

What could a guest do on a daytrip when staying at the hotel? This is what the brochure recommended:

> *The drives in the vicinity are superb. Not only is the scenery in this region of a continued delight to the eye, but the excellent roads make even the longest drives thoroughly enjoyable by carriage or automobile. The points of interest in the vicinity are numerous and include Fair Haven, Poultney, Lake St.*

The main entrance to the Glenwood Hotel as seen from land. *Courtesy of Donald H. Thompson.*

> *Catherine, the famous Middletown Springs, Hyde Manor, Carver Falls, West Rutland—where the largest marble quarries and mills in the world are to be seen—Killington Peak and Bird's-eye. A first-class livery is run in connection with the hotel for the accommodations of the guests.*

The leaflet also described Wyatt:

> *The proprietor begs to call attention to his long experience in hotel management, and to assure the public that nothing will be left undone to make the Glenwood one of the most popular and attractive summer hotels in the country.*

What is most shocking about the flyer is the second to last line: "No Hebrews Entertained."

There are two ways to view this comment in hindsight. The nature of the statement has to be recognized for what it is: a discriminatory slight against Jewish visitors. However, there may be a more benign explanation. All the guests would have expected to eat in the hotel's dining room, and it may have been a way for the management to state that it would not go to the trouble of preparing kosher food. No matter what the explanation, the 1912 season would be the only season Wyatt would operate the Glenwood Hotel.

According to multiple press accounts, the Glenwood Hotel caught fire the morning of September 17, 1912, at approximately 9:30 a.m. The fire originally broke out in the kitchen area in the back of the hotel, but within hours the entire hotel had been engulfed by a roaring inferno. It was first reported that the fire started from "an overheated chimney," but it was later reported that the blaze started when the chef's wife, Mrs. Davis, had an accident making doughnuts.

William H. Wyatt's letterhead from 1912. *Courtesy of Joseph Doran.*

According to a 1912 article in the *Rutland Herald*, Mrs. Davis spilled lard while frying doughnuts. The lard, while on the floor, was ignited by hot grease that shot off from doughnuts

The Glenwood Hotel on fire the morning of September 17, 1912. Smoke can be seen rising from the center of the building. *Courtesy of Donald H. Thompson.*

deep-frying nearby. In another slightly different account from the *St. Albans Messenger*, "an overturned donut pan" caused the fire.

The fire spread from the back of the hotel into the floors above and into the front of the hotel closer to the lake.

When Mrs. Davis realized the fire was rapidly spreading, she yelled for the remaining guests to get out. Luckily, there were only two visitors at the time. The first men at the scene from the nearby cottages only had a narrow garden hose with which to fight the fire. Luckily, there was no wind that day, so when volunteer firefighters from neighboring communities arrived, they were able to save numerous buildings, including several outbuildings and cottages.

Firefighters had their work cut out for them due to the intense heat from the conflagration. They formed a bucket brigade from the lake to douse the surrounding buildings while also throwing wet blankets and carpets on the structures. The men at the end of the bucket brigade nearest to the fire "were wrapped in soaked blankets enabling them to withstand the heat long enough to empty their pails of water."

In spite of the speed with which the fire spread, volunteers were able to save a number of items from the hotel before the walls collapsed in on the basement. The piano from the dance hall, bedroom furniture and even table linens were all saved as the fire ravaged the structure. In spite of these

Another view from Lake Bomoseen of the Glenwood Hotel on fire. *Courtesy of Joseph Doran.*

few successes, the scene that day was one of complete devastation. The volunteers on the scene were no match for the intense heat, black smoke and searing flames. By 2:00 p.m. the Glenwood Hotel was in ashes.

Where was William Wyatt when the fire occurred? According to one local newspaper clipping provided by Hydeville resident Joseph Doran, "Mr. & Mrs. Wyatt left for their home in Troy last Saturday, when the hotel was closed for the season. Only caretakers were in the building when the fire was discovered about 9:30 o'clock."

Numerous press reports indicate that the Glenwood was covered by $15,000 in insurance. The question now was: would it be rebuilt?

4

Bought and Sold

It seems that the fire at the Glenwood Hotel spurred William Wyatt to seek opportunities elsewhere. Did Wyatt use the insurance money collected from the Glenwood fire toward the purchase price of the Hulett Hotel in Huletts Landing? The answer to this question is unknown, but in order to try to unravel it, it is essential to re-create a timeline that demonstrates when Wyatt first came to Huletts Landing, saw the Hulett Hotel and became interested in making a deal.

Henry W. Buckell was the owner from whom Wyatt purchased the Hulett Hotel. Buckell took over the management of the hotel in 1886, when he was only twenty-four years old, and later purchased the property from the widow of railroad executive C.W. Wentz. According to numerous accounts, Henry W. Buckell was a capable, industrious person who also served as Dresden town supervisor from 1900 to 1904.

If one works backward from the public announcement made in the local press on January 30, 1913, that Wyatt had agreed to purchase the Hulett Hotel from Buckell, it may be possible to ascertain some clues that show what happened to the Glenwood insurance proceeds. One possible clue exists in a letter written by Henry W. Buckell to his brother-in-law, Royden Barber:

Royden,

I want you to bring over Mr. Wyatt of Lake Bomoseen, Vermont Monday. He is coming to take a look at the hotel with an idea of buying it and your

A view of the front of the Hulett Hotel from approximately 1904. *Courtesy of Lance DeMuro.*

The Hulett Hotel in 1915. *U.S. Library of Congress.*

> *job is to do some talking. Tell him, as the occasion offers, that you were clerk here five years ago and that you figured out the clear profits of the season are from eight to ten thousand. That a better class of customers are coming, paying fifteen to seventeen dollars per week. That we are always crowded, etc.*

Henry W. Buckell, owner of the Hulett Hotel prior to Wyatt, sold the building to Wyatt in 1913. *Kapusinski family collection.*

Mention the store trade, six hundred dollars of candy sales alone, five hundred of cigarettes. The cottages are always filled. The bowling alleys bring in three or four hundred, and launch parties to Fort Ti three times a week at one fifty a head, in the ECHO. *Make him think it is a paying proposition all around.*

The idea is to divide the property, selling the hotel and all the land on the north side of the big brook as you come down the South Road to the Mill, and then north of Foster Brook, which runs into the bay at the store. Keeping the cottages and the land south, which are Mrs. Buckell's.

Be very careful not to mention that you are related to me. Make believe that I want to sell and that you know of parties offering me fifty thousand dollars for it, which I refused.

Say the hotel is not very elaborate, but that the location is called the best on the lake and attracts everyone here.

The hotel and about fifteen cottages are on this lot number one fifty one. Grand farm land and the best of gardens each year. Lead him to think I might sell at this time on account of Mrs. B being sick two years and poor health myself, going to the hospital, etc.

Have a slick carriage with blankets and a robe. Be judicious, don't exaggerate too much nor too little. Show him how property has doubled here in ten years and how little there is for sale. Tell him you work for me on and off and that I called you up Monday and asked you to meet him as my team was busy. Mention the State Road is coming next year.

Yours, H.W. Buckell

The problem with this clue is that it is incomplete. The letter, believed to be have been written in 1912, is undated. If it was written before the fire at the Glenwood, it would demonstrate that Wyatt was already looking for other business opportunities before the blaze struck the Glenwood. If the

The Barber family of Dresden, New York, circa 1894. Cora Barber (far left) later married Henry W. Buckell, owner of the Hulett Hotel. Cora's younger brother, Royden Barber, is standing on the far right. *Courtesy of the Whitehall Historical Society/Ray Rose.*

Royden Barber about 1915. Barber was instructed by Buckell to show Wyatt the hotel but not say anything about being Buckell's brother-in-law. *Courtesy of the Whitehall Historical Society/ Ray Rose.*

letter was written after the fire at the Glenwood, Wyatt may have calculated that the time was now opportune for him to leave Lake Bomoseen and pursue opportunities elsewhere. Ultimately, history doesn't tell us because Buckell didn't date the letter.

In a typed letter found in the Whitehall Historical Association archives, Betty Buckell, Henry's daughter-in-law, adds a slightly more humorous account:

> *Royden claimed that he was the one responsible for Henry marrying Cora, but then he claimed he was responsible for everything good happening up there anyway...You also know the story about how he sold the old Hulett House to Wyatt and that he was still waiting, till his dying day for* [his] *$1,500 commission.*

Was the insurance money collected from the Glenwood fire used to purchase the Hulett Hotel? The only thing known for sure is that the fire at the Glenwood Hotel occurred on September 17, 1912, and the public announcement of Wyatt buying the Hulett Hotel was published on January 30, 1913. Sometime prior to that announcement, Wyatt met with Buckell and saw the property, and they ultimately concluded the deal. Royden Barber, Buckell's brother-in-law, was the person who originally showed Wyatt the property.

As for the Glenwood Hotel, it was never rebuilt.

5

The Huletts Landing Fire

After purchasing the Hulett Hotel from Henry Buckell, Wyatt had three solid, profitable years. Many accounts state that Wyatt, during these years, invested in improving the hotel and grounds. The summer of 1915 was Wyatt's third season as owner of the Hulett Hotel.

Much of the conversation in Huletts Landing that year was about the German sinking of the British ocean liner *Lusitania* off the Irish coast on May 7, 1915, killing nearly 1,200 people. The words "Zeppelin raids" and "Dardanelles campaign" would have been heard by many guests during the summer of 1915 as hostilities in Europe began to intensify. As the start of World War I consumed the world, Huletts Landing was viewed as a safe and peaceful place to go on a summer vacation. For Wyatt, it was a good year to own a hotel in upstate New York.

After Labor Day, the tourists departed, the days became shorter and off-season work began. Contractors were hired to work on improving the septic system for the hotel.

November 14, 1915, was a Sunday. The *Glens Falls Times and Messenger* reported that the night had been fair and slightly colder. The weather that morning was clear and cool. Sunday was considered a day of rest, and most people would have been attending morning church services. By midmorning, there were only two people on the grounds of the Hulett Hotel. They were the caretaker, Willis Foster, who was expecting a quiet day, and Mr. Adelbert Buckell, the brother of Henry W. Buckell, the former owner.

The last known photograph of the first Hulett Hotel before the fire on November 14, 1915. *Courtesy of Lance DeMuro.*

Foster was born on November 16, 1857, in Illinois and came to Dresden with his parents while still a boy. He had lived most of his fifty-seven years in Huletts Landing and would celebrate his fifty-eighth birthday in two days' time. He lived with his wife in the hotel during the off-season, doing routine maintenance and watching over the hotel and grounds. During the summer, he was the assistant postmaster of Huletts Landing. That morning, Foster's wife had departed to attend Sunday church services. The summer tourist season was over, and the mid-autumn weather foreshadowed winter's approach.

Adelbert Buckell was visiting his brother, Henry, who still owned other property in Huletts Landing after selling the hotel to Wyatt. Adelbert was the local teamster, responsible for driving a team of horses and a wagon over the mountain with supplies.

The day would soon change things forever. Willis Foster smelled the smoke first and then spotted the fire coming from the hotel's roofline near the main chimney. By the time he saw it, it was already a blazing inferno. Because it raged from the upper floors near the center of the building out toward the ends, it appeared to have started near the chimney. When Foster discovered it, it had reached such proportions that it was impossible to put out. It burned fast and quick; it burned hot. He and Adelbert Buckell were alone by all accounts. Upon discovering that the fire could not be brought under control, the shock must have been instantaneous. What greater regret could the lonely caretaker have felt than to see the historic structure go up in flames on his watch?

Willis C. Foster (sitting, far left), caretaker of the Hulett Hotel, was the first to see the fire. He is seen here with his six brothers about 1910. *Courtesy of Mike Foster and Kathy Huntington.*

The conflagration was so great that it was spotted almost immediately from across the lake. The fire did not stop at the hotel, either, quickly jumping to three nearby cottages as well. The *Ticonderoga Sentinel* reported:

> *The small number of men who crossed the lake to assist* [Willis Foster] *could do but little more than stand idly by and watch the flames eat up the buildings. From the hotel the fire spread to the three nearest cottages, all of the buildings burning like tinder. There are twenty cottages in connection with the hotel and for a time it was feared that all or most of them might go up in smoke, but, fortunately, the fire was confined to the three nearest the hotel.*

Firefighting in those days was primitive in comparison to the modern fire departments that exist in almost every municipality today. Even a modest volunteer fire department in a rural town today, equipped with modern trucks, hoses and wireless radios, would have been considered science fiction during that time period. In the years from 1912 to 1920, motorized fire apparatus were just beginning to debut in numerous cities across the United States. Firefighting during that time still relied on horse-drawn carriages and

Adelbert Buckell (standing, far right), brother of Henry W. Buckell, pictured in 1905 with other hotel employees. He was nearby when the Hulett Hotel caught fire. *Kapusinski family collection.*

bucket brigades. The distance between fire companies in almost every major city was predicated on how far a fire horse could pull its apparatus while running at full speed.

The goal of those responding to a fire was to prevent it from spreading. It was usually a forgone conclusion that the structure being consumed was a lost cause. Many of the major hotels in the Adirondacks of upstate New York met their ends in tragic fires. The most basic problem with firefighting in 1915 was that fire codes were almost nonexistent. The structures were built with wood and nails, and very little thought went into constructing a building that could withstand a fire.

Imagine the poor caretaker, Foster, discovering the roaring inferno in a deserted, out-of-the-way place like Huletts Landing, with little immediate assistance from neighbors and no fire company to respond. No wonder men rowed from across the lake to help. Those neighbors from the lake's opposite shore knew when they saw flames that their assistance would be needed.

Accounts are few and unclear, but one of the men present, either Foster or Buckell, galloped off on horseback to round up assistance throughout the hamlet. A nearby neighbor, Cora Phillips, was one of the first people to arrive after she saw the thick black smoke. What a scene it must have been, with neighbors rowing across the lake and either Foster or Buckell doing his best

Firemen's convention in Whitehall, New York, in 1897. When fire struck, it was usually a forgone conclusion that the entire structure would be destroyed. *Courtesy of the Whitehall Historical Society/Ray Rose.*

imitation of Paul Revere, riding through the town shouting for help. Anyone seeing the smoke or hearing the cries would have immediately stopped what he was doing and run to the fire. All the while, the conflagration ate more of the hotel, spewing black smoke and bright orange flames high into the air.

The fire worked its way out of the center of the second floor, and within a half an hour, the west end of the hotel was completely gone. Because the day was somewhat windy, the fire easily jumped to nearby cottages.

Soon, more neighbors arrived and formed a bucket brigade to stop the fire's spread. They were fortunate that the wind died down and the fire was prevented from spreading to additional cottages.

Within only two hours, the fire had completely destroyed the Hulett Hotel and three nearby cottages. Charred cinders and ashes were all that were left of the hotel. People from throughout the landing and the town of Dresden had come to look for themselves. The brick fireplace, remnants of the scarred chimney and some building supports were all that remained.

The first question everyone wanted answered was: how had the fire actually started? Adelbert Buckell claimed that the fire broke through the roof near the defective chimney, which his brother Henry had had problems with in the past.

The original "Idle Hour" cottage was one of the summer cottages that burned down the day of the Hulett Hotel fire. *Kapusinski family collection.*

Years before, Henry Buckell had cut a pipe hole in the chimney in the second floor to make it possible to set up a stove. The chimney led to the roof from a fireplace, and the chimney burned the woodwork in the attic. Henry Buckell said that when he sold the property to Wyatt, he warned him that the chimney was dangerous. Thus, the defective chimney seemed to be the most likely cause of the fire.

The mystery of the Lincoln photographs now begins because the earliest picture attached to the Lincoln illustration shows the devastation of the fire. While it appears to have been taken one to two months after the fire, as the new hotel was being built, the photograph captures the remnants of the fireplace and chimney where the fire was believed to have started.

The next question was: where was William Wyatt when the fire started? Wyatt was not in Huletts Landing at the time of the fire but at his home in Troy. The entire Wyatt family, as was their custom, had left Huletts to spend time at their other properties in Troy and Saratoga. Wyatt's health was poor; he had been experiencing heart problems. Sadly, the one newspaper account that documented his reaction when he learned about the fire was scanned to microfilm with that paragraph omitted.

However, numerous press accounts of the fire were instantaneous, with stories appearing in multiple newspapers over the succeeding days.

The *Glens Falls Times and Messenger* of November 15, 1915, screamed on page one: "Lake George Scene of $50,000 Conflagration." Two subtitles followed this headline: "Fire Attributed to Defective Chimney Destroys

The remains of the Hulett Hotel are still visible as the new hotel takes shape in the background. The "x" marks the location of the fireplace and chimney where the fire was believed to have started. *Lincoln illustration #1.*

Hulett House and Three Cottages—Resort Probably Will be Rebuilt" and "Loss Partly Covered by Insurance."

The good news of the day was that no one was killed or physically injured in the fire. All the volunteers responding to the scene were unhurt, and the only harm sustained was the property damage. Willis Foster, the faithful caretaker, received credit from the neighbors for alerting the town and averting an even greater tragedy. The fact that the hotel was closed for the season seems to have prevented anyone from being killed.

Fortunately, the hotel was covered by fire insurance, negating much of the property loss. During that time, for large structures like a hotel, the risk was spread among numerous underwriters. The Hulett Hotel was insured by thirteen different insurance companies, limiting their risk to $3,000 each. Unlike today, when having this many policies on one structure might raise suspicion, this was the normal course of business in 1915, when fires were more prevalent.

Theoretically, this meant that the Hulett Hotel was insured for a total of $39,000. However, because of the structure and the terms of the policies that apportioned the risk, Wyatt was reported to have received $37,000.

In one final interesting coincidence that would only become significant later, the *Whitehall Chronicle*, the weekly newspaper from the nearby town of Whitehall, reported on the Hulett Hotel fire in its edition of November 19.

The article, "Fire Destroys Property," was displayed in a column immediately to the left of an article on the same page in which the election results for the Washington County district attorney's election were published. A young Republican candidate named Wyman S. Bascom received 6,935 votes, while his three opponents totaled only 4,461 votes combined.

On the day that the readers of the *Whitehall Chronicle* were learning about the Hulett Hotel fire, they were also reading that a new district attorney had been elected. This would be the first of many times that the name of the new district attorney and the events surrounding the Huletts Landing fire would appear together in the newspaper.

6
The New Hulett Hotel

Only four days after the Hulett Hotel fire, the *Ticonderoga Sentinel* of November 18, 1917, reported that Wyatt planned on rebuilding quickly: "Mr. Wyatt stated that he will probably replace the burned hotel with a fireproof structure."

This quick decision to rebuild was necessary because the fire happened in November, leaving a little more than seven months before the next summer tourist season would begin. Waiting would be costly to Wyatt; he would lose the income from the upcoming season if a new hotel were not finished by the summer of 1916.

Newspaper accounts estimated the loss caused by the fire at $50,000, but the fire insurance that Wyatt carried totaled only $37,000. In order to rebuild the hotel, he would need to borrow money.

Wyatt moved quickly, contacting the First National Bank of Glens Falls for a loan. Wyatt stated later that the bank originally denied his loan request, so he made a direct appeal to Byron Lapham, the president of the bank. According to Wyatt, Lapham called a meeting of the bank directors, and they decided to loan Wyatt some of the money if the Finch-Pruyn Paper Company would loan him the rest. Wyatt, using contacts he had amassed over his years in business, was able to secure the additional funds from Finch-Pruyn. Wyatt signed a mortgage of $30,000, with the new hotel as collateral, and got to work immediately building the new hotel.

The concept of applying for permits and complying with building codes was nonexistent in Washington County in 1917. If a person owned the land, he could build on it. The hotel that burned down had been a series of additions and modifications that had taken shape over a number of years, but the new hotel would have to be constructed rapidly. Building a fireproof structure was ultimately not undertaken due to the cost and length of time needed to complete it.

However, a team of carpenters had to be assembled, and lumber had to be acquired and transported into Huletts Landing quickly.

The construction of the new hotel was begun under the direction of Schenectady, New York contractor George B. Preston. A 1905 advertisement described the George B. Preston Construction Company as follows:

> *G.B. Preston, president. Carpenters and builders. Reinforced concrete work and block manufacturers. One of the most enterprising and popular firms in their line in the city. Their work throughout the city is evidence of the reliability of the firm, and the fact that the same is appreciated is noted through accumulating orders. Their motto: First class workmanship and material at the lowest possible figures consistent with same.*

Preston was familiar with building large-scale projects and could assemble the large number of tradesmen necessary for the undertaking.

Instead of rebuilding the new hotel on the exact site of the previous hotel, a higher location, overlooking the tennis courts, was chosen nearby. This was done for two reasons: to have gravity aid the sewage disposal process and to give the rooms a better view of Lake George. The former hotel had been located at a lower elevation, on soil that was less conducive for drainage.

The design had a long rectangular layout, ample verandas, gabled pavilions and a center cupola. The book *Resort Hotels of the Adirondacks* describes the new hotel's architecture:

> *That it was a product of its time, however, was evident in the pronounced, overhanging roof eaves, low-pitched roofs, and heavy, triangular eaves brackets commonly associated with the then popular Arts and Crafts, or Craftsman, style.*

The vast majority of the lumber for the new hotel was cut at a Clemon's sawmill or brought through the railroad station there. Clemons is a small hamlet also in Dresden that sits on the opposite side of the mountain from Huletts Landing. With the steamship not operating in the winter, the only route into Huletts was over the mountain.

Lumber to construct the new Hulett Hotel crosses the top of the mountain on horse-drawn sleds during the winter of 1915–16. *Lincoln illustration #2.*

The remaining Lincoln photographs all capture, in varying degrees, the construction of the new hotel. The next photograph in the sequence of pictures attached to the Lincoln illustration shows two horse-drawn sleds, with two horses pulling each sled, crossing the top of the mountain in winter on their way to Huletts. Each sled is piled high with lumber. The bells, which can be seen in front of one horse, let people know the sleds were approaching.

What is interesting to note about the Lincoln photographs is that the locations where the pictures were taken are all still readily identifiable today. Picture two was taken on the flat part of the top of the mountain; pictures three, four and five were taken as the road on top of the mountain begins to descend into Huletts Landing. It appears that the photographer stood in one place as the two sleds went by.

The pictures of wood being pulled over the mountain on horse-drawn sleds during the winter of 1915–16 illustrate that the age of the automobile had not yet fully begun. The assembly line style of mass production of cars with interchangeable parts was greatly expanded by Henry Ford beginning in 1914. As a result, Ford's cars came off the line in fifteen-minute intervals, much faster than previous methods, increasing productivity eightfold while using less manpower. The assembly line production of Ford's Model T was so successful that, in 1914, an assembly line worker could buy a Model

Driving heavy sleds of lumber down the mountain was a tricky proposition. The deep snow could indicate a timeline of December 1915 to February 1916. *Lincoln illustration #3.*

T with only four months' pay. However, throughout the early part of the twentieth century, automobiles, while becoming more common, were still a novelty throughout the country.

Because the construction of the new hotel was undertaken so quickly, there was no basement installed. The site itself was rocky, so the base of the hotel was constructed on footings, or piers, which were anchored directly to the rock ledge. Lincoln photograph six also captures Burgess Island clearly in the background.

Conflicting reports exist regarding the naming of this state-owned camping island. Some suggest it was named for the prominent John Burgess family. Others claim its name stems from an old British word for town leader. One report that Thomas Reeves Lord makes reference to in the book *Stories of Lake George Fact and Fancy* claims that during the Revolution, an American sympathetic to the British cause supplied information about the colonists' movements to English commanders by leaving messages on the island. The British considered the spy a faithful citizen, or burgess, of the Crown.

The pictures also illustrate how large-scale projects were undertaken during that time period. Heavy equipment was almost nonexistent during construction projects from that age. When the rocky outcropping interfered with the new hotel's construction, workers used pickaxes to break the stone apart. No hydraulic backhoes or earth-moving equipment were available

Horses haul the wood for the new hotel over the mountain. *Lincoln illustration #4.*

The amount of lumber quickly brought in to construct the new hotel may have led to rumors about the origin of the fire. *Lincoln illustration #5.*

to assist them. Workers toiled doing heavy labor under what would be considered backbreaking conditions today. Workers can be seen in the photograph on the next page using pickaxes and shovels to clear by hand debris that was interfering with the ground floor.

Wood is piled high as constructions begins. The surrounding cottages and Burgess Island can be seen clearly. *Lincoln illustration #6.*

Before the first floor of the new hotel could be put in place, workers used pickaxes and shovels to clear away parts of the ground. *Lincoln illustration #7.*

The tradesmen hired by Preston worked throughout the winter dressed in warm clothes. Throughout their time on the job, they were exposed to biting cold. The leather boots they wore got stiff in the cold, making walking hard by limiting the range of motion in their ankles. Also, their toes would get

Workmen laying the first floor of the hotel. The man on the right is wearing a carpenter's vest and carrying a hammer. The original employee dormitory can be seen behind them. *Lincoln illustration #8.*

cold, damp and numb, making climbing ladders and avoiding objects more difficult. Today, laborers would be brought onto a job site daily and would have trailers or temporary shelter in which to rest periodically. Preston's men were housed in nearby cottages or rented rooms and worked twelve- to fifteen-hour days exposed to the elements. They were known for their hard work and skill. They knew what was expected of them and were undaunted by the demands of the job.

While some of the Lincoln photographs illustrate more interesting scenes, they also, in many cases, document construction and framing techniques used during construction. The new hotel was built with a method called platform framing.

In this method, the floors, walls and roof of a framed structure are created by nailing consistently sized dimensional lumber (two-by-four, two-by-six, etc.) at regular spacing, forming walls and floors. The floor, or the platform, is made up of joists (usually two-by-eight, two-by-ten or two-by-twelve, depending on the span) that sit on supporting walls, beams or girders. The floor joists are spaced at regular intervals and covered with a plywood subfloor. The floors, walls and roof are typically made torsionally stable with the installation of a plywood wood skin referred to as sheathing.

Many of the Lincoln photographs illustrate the floors being built on successive platforms, the nailing of beams and sheathing being attached.

The supports can be seen under the back of the new hotel. *Lincoln illustration #9.*

Slowly, the new hotel began to take shape. As it rose high into the air, townspeople stopped to view the construction. It was much higher than the previous hotel and offered commanding views of Lake George. Wyatt was intimately involved in the construction process, reviewing architectural plans, meeting with Preston's foremen and reviewing the progress being made daily. Two of the Lincoln pictures capture him on the job site.

The size of the building was impressive, but the overall design was quite simple. Shaped like a square letter "u," the front faced southwest across the lake, in the direction of the prevailing winds, with the wings of the building extending out, like buttresses, behind it. The exterior sheathing had been freshly cut as it was put on. It would only be painted later, once the structure was completed.

The interior courtyard, with one side open, was located away from the lake on the mountain side of the hotel. There was an entrance from the courtyard for the staff to enter from the nearby employee dormitory.

Opposite, top: William H. Wyatt (left) stands with another man reviewing a document while construction continues behind them. *Lincoln illustration #10.*

Opposite, middle: William H. Wyatt stands on what would become the front porch of the new hotel. *Lincoln illustration #11.*

Opposite, bottom: The third floor of the new Hulett Hotel can be seen taking shape. *Lincoln illustration #12.*

The front of the hotel under construction. *Lincoln illustration #13.*

Workmen adding the third floor to the hotel. The center of the photograph shows stockpiled wood almost two stories high. *Lincoln illustration #14.*

William Wyatt's son, Arthur, was also captured on film. Born in 1884, Arthur Wyatt was thirty-two years old in 1916. He had rented the old hotel from his father for a flat fee and in return made whatever profit the building produced. He was also actively involved in the construction process because

Arthur Wyatt, William H. Wyatt's son, stands facing the camera. The front of the building is now more complete and higher. *Lincoln illustration #15.*

This photograph, one of the most badly damaged of those attached to the Lincoln illustration, shows the roof being installed and the window cutouts being made. *Lincoln illustration #16.*

he was the person in charge of the day-to-day management of the Hulett Hotel. One poor-quality image captures Arthur Wyatt wearing a winter fur hat and looking directly at the photographer.

While the quality of some of the photographs attached to the Lincoln illustration are poor, they nevertheless capture the grandeur of the hotel as

This photograph shows the new hotel nearing completion. While the Hulett Hotel was larger, the similarities to the Glenwood Hotel are evident. *Lincoln illustration #17.*

The woman pictured is believed to be Ella Haynes Wyatt, William H. Wyatt's wife. The ice on the lake appears thin, indicating that the photograph was taken near winter's end in 1916. *Lincoln illustration #18.*

it rose over the surrounding landscape. The area that once contained the old hotel, which had burned, was manicured into gardens that blended into the grounds of the new hotel.

William Wyatt's wife, Ella Haynes Wyatt, is also apparently pictured. She co-owned the property with her husband but was not reported to be active in the business. Little is known about her, other than she was also born in 1861, the same year as her husband. She is seen in one of the later photographs as the porch nears completion and the ice melts on the lake behind her.

As winter turned into spring, the new hotel neared completion. The center cupola was finished, and the roof was shingled. The building was painted white with dark green lattice adorning the base. The inside was classic white with elegant wood molding. The final touch was for the American flag to be hoisted high above the center of the building to fly in the ever-present breeze coming from Lake George. The flag was raised and lowered through a square hole in the cupola through which the flagpole was centered. The Wyatts, the builders and many townspeople were present as the flag was lifted for the first time above the new structure.

The *Lake George Mirror* of June 24, 1916, summed up the herculean task that had been accomplished during the previous seven months while also describing the new facilities:

The completed new Hulett Hotel days before it opened in June 1916. The canoes and early spring foliage indicate that this picture was probably taken in May 1916. This is the last picture attached to the back of the Lincoln illustration. *Lincoln illustration #19.*

During all of last winter teams and men amounting to a small army were rushing to complete what is to become one of the most popular caravansaries of the north. Mindful of the importance of correct sanitation and determined to have a resort that would be beyond adverse criticism of any sort Messrs. Wyatt went to great expense and unlimited effort in discovering what would be the best solution of the sanitary problem. The site of the new house is on considerable elevation northeast of the old hotel. From the vantage point of this rise it became possible to inaugurate a system of chemical sewage disposal which is one of the most effective in existence. All the waste from the kitchen, laundry and bath rooms is taken care of that there is nothing on the premises which can in any way breed disease or give offensive odor, neither does any of the sewage reach any stream or swamp where it might be carried into the waters of Lake George…

The new hotel has one hundred sleeping rooms, most of which command a broad view of the lake. It is a three story house with two large ells; and its construction makes the view from every room most pleasing. There are numerous suites with private baths, and every room has hot and cold water. The rooms are well furnished with modern furniture, and in keeping with the color scheme throughout, are done in white and natural wood. An electric

A postcard of the new Hulett Hotel as it appeared when it opened in 1916. *(Kapusinski Family Collection)*

> *light plant of ample capacity has been installed by Seymour Taylor of Glens Falls...In all the public rooms the semi-indirect fixtures are an ample supply of well diffused light.*
>
> *The dining room is very large and contains a seating capacity for two hundred thirty people. On the wide veranda adjoining is being arranged an out of door dining room which will be screened in and used for those who prefer it.*

In an advertisement entitled "New Picturesque Huletts" in the July 1, 1916 edition of the *Lake George Mirror*, the hotel's new attributes were highly touted: "Running water in every room. Electric lights. Private baths." The construction had finished on time, and Wyatt had seemingly overcome the tragedy of the fire. Paying customers would be returning for a new summer. Thus, as the 1916 season opened, the future seemed bright for the Wyatts and their new hotel in Huletts Landing.

That would soon change. There were ominous times ahead.

7

Suspicion

In a small town, everyone talks. The extremely quick rebuilding of the Hulett Hotel is what most likely led to the rumors. Some of these persist even today. The one that seems to have the most staying power is the allegation that the wood for the new hotel had already been cut when the fire happened, leading to the conclusion that the fire must have been deliberately set and that there was a plan in place to build a new hotel even before the fire started. Did these rumors eventually reach the authorities? Is this how the arson investigation started? It is impossible to say almost one hundred years after the fact. However, the investigation started somewhere, and there are many alleged motives for the fire being deliberately set. In an attempt to focus on why charges were brought, a systematic analysis of the reasons arson may have been committed need to be examined.

The amount of wood that was needed to construct the new hotel was obviously large. All of the accounts, which are corroborated by some of the Lincoln photographs, indicate that the wood for the new hotel was brought over the mountain in the winter by teams of horse-drawn carriages. Let us confront one known rumor head-on: that the wood for the new hotel had already been cut when the fire happened, thus proving arson had occurred.

For argument's sake, say that a local house burns down, and the owner, wishing to rebuild, goes to the nearest lumberyard. He purchases the lumber to construct a new house. Is he guilty of arson because the lumber has already been cut? Obviously not because any lumberyard would have

The flag was raised above the new Hulett Hotel days before it opened. *Courtesy of the Old Fort House Museum/ Fort Edward Historical Association.*

some supply of wood on hand and would be able to order or cut more fairly rapidly as construction progressed and further supplies were needed.

What likely started this particular rumor was the large quantity of wood that was needed to rebuild the hotel. Building a large hotel constructed entirely of wood is more complex than constructing a modestly sized house. The local citizenry would have been keenly aware of this distinction as they saw the lumber coming over the mountain. However, does that prove arson? Certainly not. The nearest sawmill would have had some wood on hand and could have cut or obtained more fairly rapidly. During the winter of 1915–16, the Adirondacks were at their zenith of producing lumber.

Were the rumors enough to start an investigation? Did the amount of wood and the fast rebuilding of the hotel raise suspicion among the authorities? History does not tell us, but the amount of wood needed to rebuild and the ready supply on hand may have started people, including the authorities, talking. What could have been the motive that prompted an arsonist in this case?

The Sewage Disposal System

In February 1916, an article appeared in the *Ticonderoga Sentinel* that alleged serious pollution of Lake George emanating from the Hulett Hotel and surrounding cottages during the previous summer season of 1915. Entitled "Complain of Lake George Pollution: Commission Asked to Stop Discharge of Waste into Lake," it states clearly that the hotel was being investigated for causing pollutants to reach Lake George:

> *Pollution of summer resort waters to a dangerous extent has recently been shown by complaints filed with the Conservation Commission and the State Board of Health. Both are the result of investigations and both show a criminal carelessness on the part of corporations and others whose best interests require that the waters they defile should be kept absolutely pure. The complaints now made public are confined to Saratoga Lake and Lake George.*
>
> *Lake George, the source of water supply for thousands of summer visitors and campers yearly, supposedly guarded by the strictest laws, was towards the close of the past season carefully inspected by a committee of the association which looks after its interests, headed by the eminent Dr. Jacobi of New York, aided by Engineer C.A. Howland of the State Department of Health. To the consternation of the association, the investigation resulted in the discovery of so many abuses in the form of drainage and other pollutions of the waters of the lake that it was recommended officially that action be taken at once "To abate the unsanitary conditions which not only create nuisances, but also endanger public health through contamination of water supplies derived from the lake."*
>
> *The report of the engineer mentions specifically the Algonquin hotel, the hotel and cottages at Huletts Landing, Cleverdale and the steamboats of the Lake George Steamboat Co. as offenders.*

Contemporaneous newspaper articles confirm that the Hulett Hotel was being investigated for sewage pollution. The new hotel was also constructed at a different, higher elevation to supposedly help the effluent drain more rapidly through a series of gravity-fed pipes. In the trial that followed, a good bit of testimony revolved around the sewage disposal system for the hotel. Was the faulty sewage disposal system the motive for arson?

The Age of the Hotel

All buildings have a life span. After time, all structures need work. High traffic buildings like hotels usually need more work, not less; roofs need to be replaced, plumbing breaks, repairs have to be made.

There is conflicting historical information about exactly when the original Hulett Hotel was built. However, it seems that the growth of the hotel was a work in progress. According to the book *No Dull Days at Huletts* by Betty Buckell, the Hulett family first started taking in tourists in 1873. The occupancy that first season was 30. In 1883, it jumped to 100; by 1889, it stood at 125; by 1901, it had reached 200; and by 1908, occupancy was 250. This steady upward movement of seasonal tourism indicates that the hotel was a series of buildings and additions that grew as tourism expanded. Using this timeline, by 1915, parts of the building would have been forty-two years or older.

A building reaching that age would be nearing the time when considerable work would be needed. Was the cost of that work prohibitive? The one significant problem with the hotel's age being a possible motive for arson is that there are numerous accounts that Wyatt invested heavily in repairs and maintenance before the fire. This would seem to be evidence that would exonerate Wyatt. Why would Wyatt supposedly burn down a building in which he had invested significant amounts of capital to improve? One would surmise that if the age and poor condition of the structure itself were the motive for arson, the alleged criminal deed would have been carried out before a significant investment was made in improving it.

Wyatt's Previous Ownership of the Glenwood Hotel

After fire destroyed the Glenwood Hotel in Lake Bomoseen in 1912, there was never any hint of wrongdoing or foul play. However, there were certain similarities between the Glenwood fire on Lake Bomoseen and the Hulett Hotel fire on Lake George:

1) Both fires happened after the Labor Day holiday officially ended the summer tourist season and at approximately the same time of day. The

Glenwood Hotel burned on September 17, 1912, and the fire is reported to have started at approximately 9:30 a.m. The Hulett Hotel fire occurred on November 14, 1915, and was first spotted at roughly 9:30 or 10:00 a.m.

2) There were no guests staying in the Hulett Hotel on the day of the fire. The Glenwood Hotel had two paying customers the day it burned down.

3) Both fires started in the center of the building and burned outward, making them both almost impossible to extinguish. The Glenwood fire started in the kitchen, at the center of the first floor, near the chimney. The Hulett fire began in the center of the structure, close to the chimney, on the second floor.

4) Wyatt was not at either hotel when the fires struck. Hired caretakers were on hand at each hotel on the day of the respective fires.

William H. Wyatt had the misfortune of owning two hotels that burned down within a little more than three years' time. Did the similarities of the two fires draw the authorities' suspicion?

Something certainly got their attention.

All of these factors may have contributed to the start of the investigation that ended with a formal indictment. The trial that followed centered on the actual testimony of individuals. Someone would testify that William H. Wyatt had paid him to start the Huletts Landing fire.

8
The Arrests

The arrests of William H. Wyatt and John D. Sharpe were big news throughout the Adirondacks, with stories appearing on the front page of both the *Ticonderoga Sentinel* and the *Adirondack Record* about their detention. While the accounts in these two newspapers differ slightly, they paint a picture of what transpired during the day of their arrests.

According to the *Ticonderoga Sentinel*, William H. Wyatt was arrested on Sunday, April 29, 1917, by a Troy detective at the Trojan Hotel and turned over to Sheriff McClarty, of Washington County, who had a bench warrant for his arrest. John D. Sharpe stated later that he was arrested on April 25, 1917, while he was exercising horses in Round Lake and was then transported to Hudson Falls.

According to the code of criminal procedure for the state of New York used in 1917, a bench warrant for a felony would be issued by the clerk of court on the application of the district attorney after an indictment had been issued. It can be assumed that the police officers who arrested Wyatt and Sharpe carried a bench warrant that most likely followed the basic 1917 form:

> *County of Washington:*
> *In the name of the People of the State of New York:*
>
> *To any peace officer in this state. An Indictment having been found on the day of April 23, 1917, in the county court of the county of Washington*

Wyatt's Trojan Hotel in Troy as it looks today. The sign still hangs in front. Wyatt was arrested at the Trojan Hotel on the charge of arson in the second degree. *Kapusinski family collection.*

charging [William H. Wyatt/John D. Sharpe] *with the crime of arson in the second degree.*

You are, therefore, commanded forthwith to arrest the above named [William H. Wyatt/John D. Sharpe] *and bring him before the county court of Washington, to answer the indictment; or if the court have adjourned for the term, that you deliver him into the custody of the sheriff of the county of Washington.*

By order of the court.

A bench warrant would have been signed and dated by the clerk of the court.

The *Ticonderoga Sentinel* account states that Wyatt was indicted by a Washington County grand jury for arson in the second degree and charged with hiring John D. Sharpe to burn the Hulett Hotel. He was arraigned before Justice George R. Salisbury on April 30, 1917, pleaded not guilty and was released on $10,000 bail. Wyatt denied any knowledge of the origin of the fire. Sharpe was indicted on the same charge and was also arraigned but failed to furnish $2,500 bail.

The *Glens Falls Times and Messenger* reported that Wyatt's bonds were supplied by Norman H. Sherry, a prominent wholesale grocer from Troy, and Lake George supervisor Edwin J. Worden, proprietor of the Worden and Arlington Hotels in Lake George Village. Wyatt was represented at his arraignment by attorney H.W. Williams.

Both the *Ticonderoga Sentinel* and the *Adirondack Record* make reference to the fact that the Washington County grand jury returned its indictment on the previous Monday, which would have been April 23, 1917.

The story in the *Adirondack Record* has the same set of facts as the *Ticonderoga Sentinel* article except that it recounts that Wyatt was arrested in Troy on Saturday, April 28, 1917. It also states that Wyatt was the proprietor of the Trojan Hotel in Troy, while the *Sentinel* article is a bit more vague, describing Mr. Wyatt as running "a hotel in Troy." The *Sentinel* article also refers to a former Wyatt hotel when calling him the recent "proprietor of the New Columbian hotel in Saratoga."

It is clearly apparent that Wyatt's name is the more prominent of the two, and the press clearly highlighted his involvement. While news traveled a bit slower in the early part of the twentieth century, the newspapers realized that this was indeed an interesting story.

What is left unsaid by all accounts is that Sharpe's inability to furnish bail meant that he would have been immediately incarcerated in the Washington County jail. Historical accounts from that period indicate that there were very few prisoners in the jail at the time. In March 1917, the month prior to the arrests, there were only twenty prisoners held captive there. This was recorded as the smallest number of inmates for many years.

Sheriff Robert J. McClarty was known locally for taking a tough approach to vagrants and tramps, as they were referred to in the local press. When the local sheriff deemed it necessary, McClarty had the authority to send an inmate to the Albany Penitentiary for further jail time.

The Albany Penitentiary was considered a model reform facility. A local sheriff could transfer inmates there if he decided to. *New York Public Library.*

The Albany Penitentiary's success as a model reformatory of the day is historically documented. The state legislature passed act after act permitting other counties to use the facility. The growing roster of inmates was put to work caning chairs and making inexpensive shoes. It seems McClarty's reputation was not considered overly hospitable to the occupants of his jail. Wyatt's ability to post bond would mean that he would be able to prepare for his trial with his attorneys as a free man.

The 1917 code of criminal procedure for the state of New York defined arson in the second degree in section 222 as a person who:

> *1. Commits an act of burning in the day time, which, if committed in the night time, would be arson in the first degree; or*
>
> *2. Willfully burns, or sets on fire, in the night time, a dwelling house, wherein, at the time, there is no human being; or*
>
> *3. Willfully burns, or sets on fire, in the night time, a building not inhabited but adjoining or within the curtilage of an inhabited building, in which there is, at the time, a human being, so that the inhabited building is endangered, even though it is not in fact injured by the burning; or*
>
> *4. Willfully burns, or sets on fire, in the night time, a car, vessel, or other vehicle, or a structure or building, ordinarily occupied at night by a human being, although no person is within it at the time; or,*

> *5. Willfully burns, or sets on fire, a vessel, car, or other vehicle, or a building, structure, or other erection, which is at the time insured against loss or damage by fire, with intent to prejudice or defraud the insurer thereof.*

Being found guilty of the crime of arson in the second degree carried a punishment of "imprisonment for a term not exceeding twenty-five years." Newspapers accounts of the day highlighted that the penalty for a person convicted of this charge was twenty-five years at hard labor.

The charge itself gives a glimpse of the prosecution's case against both men in so far as it differs from arson in the first degree in a few important ways. Arson in the first degree at that time was defined as:

> *A person who willfully burns, or sets on fire, in the night time*
>
> *1. A dwelling house in which there is, at the time, a human being;*
>
> *or*
>
> *2. A car, vessel, or other vehicle, or a structure or building other than a dwelling house, wherein, to the knowledge of the offender, there is, at the time, a human being.*

For the purpose of the actual charge, the time the fire was alleged to have been set is unclear, but no one was present in the hotel at the time the fire was deemed to have started. The rental cottages were considered dwelling houses, but no human beings were present at the time of the fire, as the summer season was over and the business was closed. Thus, the suspects were not charged with arson in the first degree but instead were indicted for arson in the second degree.

Only later was it learned that the grand jury threw out a second indictment for arson in the third degree against both men. The district attorney clearly had enough information to bring an indictment, but how strong was his case?

9

The Cast

Wyman S. Bascom

Washington County District Attorney

Wyman S. Bascom officially became Washington County district attorney in April 1916. He was thirty-one years old when he was sworn in.

Washington County district attorney Wyman S. Bascom in 1916. *Courtesy of the Old Fort House Museum/Fort Edward Historical Association.*

Shortly thereafter, on Friday, May 12, 1916, Bascom lost his wife in a fatal car accident. According to multiple press reports, he and his wife, along with their three children, were on their way to a fishing camp in Blue Ridge about seventeen miles north of Schroon Lake. Bascom was driving, his wife was seated beside him and their three children were in the rear seat. The oldest child was six years old, and the youngest was two. It was approximately 6:30 p.m. The travelers had gone up a steep road, which was in bad condition, when their car hit either a pothole or a large bump. The engine stalled, and the floorboards of the car were jarred loose, preventing Bascom from applying the brakes. The car rolled backward down the steep incline, picking up significant speed until it bounced over a small embankment. The car rolled over, throwing Bascom and his three children clear but pinning Mrs. Bascom underneath when it came to rest.

Although one of his knees was severely injured, Bascom hobbled to the car and tried to lift it off his trapped wife. Try as he might, Bascom was unable to move the car on his own, and his wife quickly died.

"Stunned by the realization of her death and the narrow escape of his little ones and himself, Mr. Bascom was forced to leave the children to guard the silent form of their mother while he walked to the nearest farm house two miles away and summoned assistance," the *Essex County Republican* reported.

A number of men hurried to the scene and overturned the car, releasing the body. A *Ticonderoga Sentinel* account continues to describe the horrific day:

> *Dr. Breen was summoned from Schroon Lake. His services, however, were of no avail. The woman's body showed no marks and the physician stated that he believed her spine had been fractured and that she had received internal injuries. W.W. Bruce of Blue Ridge arrived at the scene with his car shortly after the doctor and he took Mrs. Bascom's body as far as Schroon Lake, the children following in another car.*

Mrs. Ester Cowles Bascom was only thirty-two years old on the day of her tragic death.

The grief from such a sudden loss must have been debilitating for Bascom. But this was the situation he found himself in; Bascom was now a widower, alone and with three young children entirely dependent on him. He was also the new district attorney of Washington County. Did this loss cloud his judgment? Did the pain make him miss things he wouldn't have otherwise? Was he distracted? Did he blame himself? History doesn't tell. All that is known is that the accident happened, and Bascom was the district attorney of Washington County when it did.

However, this horrific memory must have still been fresh in the district attorney's mind as he began the arson trial against Wyatt a year and three days later on the afternoon of May 15, 1917.

William Sears

Attorney for the Prosecution

Today, a modern district attorney's office has a number of assistant DAs. In 1917, however, the practice was to hire local trial attorneys to assist the DA on a case-by-case basis. Numerous accounts state that Wyman Bascom was assisted in the prosecution by William Sears, a Whitehall attorney.

Mr. Sears was an experienced trial attorney who brought numerous talents to the prosecution table. But only a year earlier, he had had his own run-in with the law, being indicted by a grand jury for grand larceny on April 25, 1916.

This is how the *Ticonderoga Sentinel* described that event:

> *Attorney William Sears of Whitehall, one of the best known practitioners in the northern part of the county, was indicted for grand larceny in the second degree. He was charged with the theft of $132. The complaining witness was Elizabeth Moulton of Whitehall for whom Sears is said to have transacted legal business. It was charged that the lawyer failed to turn over to his client the amount due her. He entered a plea of not guilty and bail was fixed at $500.*

A year later, with the successful disposition of his own case behind him, Sears joined Bascom to prosecute the arson charge. It is safe to surmise that one reason Sears was retained, in addition to his formidable trial skills, was his familiarity with the site of the fire and his knowledge of the immediate area because he was from Whitehall.

Henry W. Williams

Attorney for William H. Wyatt

When William Wyatt was arrested, he was originally represented by attorney Henry W. Williams of Glens Falls. Williams was the son of Sherman Williams, a noted historian and the head of the school district. According to the 1930 census, Henry W. Williams was born in 1875, which would have made him approximately forty-two years old in 1917. He was married to Agnes B.

Williams and had one daughter, Josephine. His law office was in the Glens Falls Insurance Building.

One would expect that he was an established and capable lawyer. However, this case would be a challenge. A more experienced attorney had to be included on the defense team.

JOSEPH A. KELLOGG

Defense Counsel for Wyatt

Joseph A. Kellogg, Wyatt's defense counsel, in a campaign photograph from 1904 while running for Congress from New York. *Crandall Public Library.*

Joseph Augustus Kellogg was William Wyatt's primary trial defense attorney. To say he was connected with the political establishment would be an understatement.

Joseph A. Kellogg was born in 1865 and came from a distinguished family that traced its lineage all the way back to the original Joseph Kellogg, who was born in Braintree, England, in 1621 and was the first of the line to settle in America. Another ancestor, John Kellogg, was a soldier in the French and Indian War and later participated in the Revolution, serving at the Battle of Lexington and the capture of Dorchester Heights. Another member of the family, Elijah (or Elias) Kellogg, took part in the capture of Fort Ticonderoga with Ethan Allen and Benedict Arnold.

At the age of only twenty-six, in 1891, Joseph A. Kellogg was elected to the New York State Assembly from Washington County's Second District. In 1904, he was the Democratic nominee for U.S. representative from New

York's Fourth District. While he was unsuccessful in his bid for Congress, one newspaper account refers to him as "a brilliant lawyer."

In early 1911, he was appointed as a justice of the Supreme Court from New York's Fourth District by Governor Dix to succeed the late Justice Spencer. He was nominated by the Democratic Party to run for a full term in October 1911 but lost the general election. As a consolation, the attorney general of New York, Thomas Carmody, reappointed him first deputy attorney general. This was the position he was serving in when Governor Dix selected him to fill the unexpired term of Justice Spencer. In 1912, he would go on to serve as a delegate from New York to the Democratic Convention, where Governor Woodrow Wilson of New Jersey became the party's standard-bearer. In 1914, Kellogg, along with Attorney General Carmody and state senator George Blauvelt, formed a new Glens Falls law firm. One of their first clients was the powerful Delaware and Hudson (D&H) Railroad.

Kellogg could have become acquainted with Wyatt through this connection to the D&H Railroad. Kellogg represented the railroad in many cases that were related to its operation. For instance, he was one of the D&H's lawyers who successfully defended the company in a well-publicized 1914 case in which the estate of a Ticonderoga man alleged that the railroad was responsible for the man's death while he was crossing the tracks at the Ticonderoga station. Kellogg successfully defended the case, and the claim was dismissed. The Delaware and Hudson Railroad was the single largest carrier of tourists into the region during this period. Wyatt's customers and guests at Lake Bomoseen, and later at Huletts Landing, would be very familiar with traveling on D&H trains.

To better understand Kellogg's formidable speaking and reasoning abilities, an account from the July 6, 1916 edition of the *Ticonderoga Sentinel*, sums up his enormous persuasive abilities. Kellogg was described as giving the "principal address" at the dedication and unveiling of the Civil War Soldiers' Monument in Ticonderoga:

> *The massive granite block with its life size bronze figure of a Union infantryman, the gift of Hon. Clayton H. DeLano, that stands in the Central School park as a memorial to those Ticonderogians who fought, bled and died for the Union was formally dedicated, presented and accepted on Tuesday with exercises that were befitting the day, the gift and the giver… The principal address of the day was made by Hon. Joseph A. Kellogg of Glens Falls, a brilliant and forceful speaker, whose speech on Patriotism*

> *and Preparedness teamed with true Americanism and whose discussion of the mooted question of preparedness, scholarly in construction and the result of deep thought on this subject, was enlightening and interesting in the highest degree.*

There is no doubt that Judge Kellogg was well connected politically, as well as a gifted orator and lawyer. Wyatt had retained one of the best attorneys in the state to defend him.

ERSKINE C. ROGERS

Washington County Judge

Judge Erskine C. Rogers was the youngest county judge in New York when he was appointed in 1916. *Courtesy of the Old Fort House Museum/Fort Edward Historical Association.*

Erskine C. Rogers was the Washington County judge who presided over the case. He was born on September 17, 1878, in Sandy Hill, now Hudson Falls, the son of General James C. Rogers and Elizabeth Coleman Rogers. He was educated in public schools and completed his education at Union College and Albany Law School. He was president of his law school class and received the Amasa J. Parker Prize for high standing in his studies.

In 1908, he married Helen Annette Wakeman, daughter of Mr. and Mrs. Abraham Wakeman of Hudson Falls. Mrs. Rogers's paternal grandfather, Abram Wakeman, was at one time collector of the Port of New York and a personal friend of President Lincoln.

Rogers's first political appointment was that of attorney for the village of Sandy Hill. When Washington

County district attorney Robert O. Bascom died in May 1909, Rogers was named on June 4 as his successor by Governor Charles Evans Hughes.

He was thereafter twice elected Washington County district attorney. It was while in this office that he prosecuted the Great Meadow Prison graft cases in which state officials and prominent politicians and contractors were involved.

The story involved ten men who were indicted by a Washington County grand jury on charges of graft in connection with the construction of the Great Meadow Prison in Comstock. Among those indicted were Cornelius V. Collins, former state superintendent of prisons, and Charles S. Boland, president of the Hudson Valley Construction Company. While the state paid approximately $500,000 for the construction of the prison, the grand jury, in its indictment, found that the actual value of the work was only $200,000. Rogers's prosecution of the case garnered him significant publicity in local and statewide newspapers.

In April 1916, while still serving as district attorney, Rogers was appointed by Governor Charles S. Whitman to be Washington County judge to fill the vacancy created by the resignation of Judge C.R. Paris, who happened to be a cousin of Mr. Rogers.

Rogers was described as being hardworking, earnest and of good judicial temperament. At the time of his designation by Governor Whitman, Judge Rogers was the youngest county judge in New York State.

10

The Trial

By all accounts, the arson trial of William H. Wyatt started in the Hudson Falls courthouse on Tuesday, May 15, 1917. Each defendant had demanded a separate trial, so Wyatt's trial was the first to begin. In 1804, Washington County had become a "half-shire" county with two county seats. One courthouse was built in Sandy Hill (Hudson Falls), but another was built in Salem because General John Williams, a powerful Salem landowner and politician, influenced the county to keep one there. It wasn't until 1993 that the county seat moved to new facilities in Fort Edward, which action officially ended the half-shire system.

The Hudson Falls courtroom in 1917 was an impressive room. Today, the building contains a restaurant, and the courtroom itself is used as an entertainment hall. However, the room is still basically configured as it was in 1917. As one enters the main entrance off Main Street, double doors open into a foyer where two large, imposing staircases await. One staircase is on the left; the other is on the right. At the top of the inlaid tile stairs, one enters a second foyer-like area, and then, through another set of doors, one enters into the back of the courtroom. The ambiance of the room is different today, but the handcrafted flower molding at the edge of the thirty-foot ceiling still gives the room a powerful, respectful atmosphere. The benches where the public sat during the trial are still in their original locations today.

In the left of the courtroom (to the right of the judge) was the jury box, and to its immediate rear was the jury room where deliberations took place. As one faced the judge on the right, there was a door with a curved arch

The Washington County Courthouse in Hudson Falls as it appeared in 1917. Wyatt's arson trial took place there. *Courtesy of the Old Fort House Museum/Fort Edward Historical Association.*

The visitors' area of the courtroom, as seen today, is almost identical to how it appeared in 1917. *Kapusinski family collection.*

above it that led into the judge's chambers. A rail separated the lawyers' area from the public space, where there were benches for about two hundred observers. The courtroom was approximately sixty-five feet across and about seventy-five feet deep. It was then, and remains today, an imposing room. Above the back of the courtroom, as one enters, there is a loft with additional seating.

It was in this very room that the events of the trial unfolded. It is easy to imagine the bench with the imposing figure of the trial judge at the front of the room. Even though he was relatively young, Judge Rogers surely impressed those in the courtroom with the solemnity and significance of the occasion. In front of the bench, there were two oak tables of imposing length where counsel sat. There were several captain's chairs lined up behind the tables, and on the right side of the bench where the judge sat, there was a dais where the clerk of court sat. The visitors' area of the courtroom, by all accounts, was packed throughout the entire trail.

One of the challenges in re-creating the trial from 1917 is the loss, over time, of important documents that could aid research into the accurate portrayal of events. The trial transcript from Hudson Falls is now apparently lost to antiquity. However, the Washington County archives were able to provide some very important documents. The court calendar, notes written on the back of the original indictment by the clerk and a partial list of witnesses still exist in the county archives today and are considered public documents.

The Washington County archivist also unearthed two pages of handwritten notes made by Wyman Bascom that summarized the trial. These were not considered public records, so a request had to be made to the current district attorney to obtain them.

Even with the scarcity of primary source documentation, the record of the trial can still be re-created from secondary sources of contemporary newspaper articles. It should be noted that the news media of the day were thorough in their reporting of the events of the trial. The newspapers of the day devoted considerable coverage to the trial because it was a major story at the time.

The account of the trial that follows comes primarily from the reporting of the *Glens Falls Times and Messenger* but includes, in some instances, accounts from other newspapers or sources. Some minor inferences have to be assumed, however, based on the lack of primary source historical documents. Every attempt has been made to use the original words of the witnesses to preserve the historical record.

TUESDAY, MAY 15, 1917

The jury that was empanelled on the morning of May 15, 1917, was composed of twelve men from seven different Washington County towns. Specifically, the individual members of the jury were: Charles K. Ketchum, Fort Edward; Dewitt C. Ross, Fort Edward; Charles Wicks, Fort Edward; Harry C. Wicks, Fort Edward; Thomas Dean, Granville; Robert Hughes, Granville; Bert Wilkinson, Greenwich; George McNeal, Hebron; Albert Pettys, Jackson; Otis Barrett, Putnam; Henry Harris, Whitehall; and Charles H. Sherman, Whitehall.

The women's suffrage movement was in full swing in 1917, but it would not be ultimately successful until 1920. Thus, women were not only denied the right to vote in 1917, but they were also prevented from sitting on juries. Wyatt's and Sharpe's fates would be decided by juries of twelve white men from the small yet geographically diverse rural county.

At 1:00 p.m., following a short lunch break, Wyman S. Bascom stood and began the prosecution's case. He started by laying out what today would be considered the prosecution's theory of the case. Bascom stated that some years prior, Wyatt had operated the Glenwood Hotel on Lake Bomoseen, Vermont, and at that time he had become acquainted with John Sharpe, then of Ballston, New York, who was a horseman and at times conducted a livery business. The two had gone on to do business together, with Sharpe running a livery at Wyatt's Vermont resort.

According to the district attorney, Sharpe wrote to Wyatt in November 1915 to borrow some money. It was then that Wyatt went to Albany to see Sharpe, who was at that time running a livery business and living with a woman named Mrs. Cornelia Gries. Bascom made note of the fact that Mrs. Gries was a married woman and that Sharpe was in fact married to someone else at the time he was living with Mrs. Gries. This salacious detail was included in accounts of the case published in local newspapers of the time.

Bascom would go on to say that Wyatt talked with Sharpe and Gries and then made a proposition to Sharpe to burn the hotel. A day or two after their meeting in Albany, Sharpe and Mrs. Gries went to Clemons and registered at a Clemons hotel as Mr. James Smith and wife. They then proceeded to hire a carriage and driver to take them over the mountain to Huletts Landing, where they inspected the area. They saw the hotel and surrounding cottages and then returned to Clemons. They then left for Albany by train later that evening.

According to Bascom, the next day, Sharpe, accompanied by his employee, Wallace H. Strickler, took the train to Whitehall and walked with Strickler to Huletts Landing. They arrived at the landing at 2:00 a.m. on Sunday, November 14, 1915, and entered the hotel through an open door that had been arranged with Wyatt in advance. Bascom alleged that they walked to the second floor of the hotel until they entered a specific room, where they found a mattress soaked in oil. The mattress was partially on the floor and partially leaning against a wall. At one end of the mattress, there was a hole in the floor where there was a quantity of wood chips and paper. Sharpe arranged the paper to lead to the mattress. He then took a candle and placed it so that when it burned low enough, it would ignite the paper and mattress, both soaked with accelerant. According to the prosecution, Sharpe then lit the candle, and both he and Strickler exited the hotel and headed back to Whitehall. The building would be engulfed in flames only after they had been gone for a number of hours.

Soon after the fire, the district attorney said, Wyatt visited Albany and was seen by a disinterested witness to pay Sharpe $200. Later, Wyatt would pay other sums, amounting in total to $500. These amounts were considered quite large in 1915. According to the prosecution, Wyatt's motive for the crime was to collect insurance money on the antiquated hotel, which was having major sewage problems and was being investigated by the state health department.

District Attorney Bascom's opening statement foretold a lively trial. He finished by saying that John Sharpe, his employee Wallace Strickler and Mrs. Cornelia Gries would all be sworn in for the prosecution and testify to what he had described in his opening remarks.

The prosecution started the state's case by calling its first witness. Mrs. Laura Holcomb, wife of Charles D. Holcomb of Clemons, proprietor of a hotel there, who was sworn out of the scheduled order by the prosecution because her husband was ill. She testified that in November 1915, a man and woman registered as James Smith and wife of New York. However, the date they registered did not appear in the guest register, which was entered into evidence. Mrs. Holcomb identified Mrs. Gries, who was present in the courtroom, as the woman who had registered as Mrs. Smith. However, she could not identify Sharpe as the man who had registered as Mr. Smith.

The prosecution finished its questioning of Mrs. Holcomb by getting her to testify that a horse and carriage were hired by the man and woman (whom she knew as Mr. and Mrs. Smith), with Mr. Arthur Derby as the driver, to take them to Huletts Landing.

The Clemons train station in 1915. John D. Sharpe and Cornelia Gries testified that, while traveling under assumed names, they passed through the Clemons train station on the Friday before the Hulett Hotel fire. *Courtesy of the Old Fort House Museum/Fort Edward Historical Association.*

The defense now had its first opportunity to begin. Joseph Kellogg, Wyatt's eloquent defense attorney, who would be referred to in press accounts throughout the trial as "Judge Kellogg," now stood and began his questioning of Mrs. Holcomb. On cross-examination, Mrs. Holcomb admitted that she did not recall seeing the man and woman in question actually register. She stated that the man and woman who registered as Smith were two of eleven people who had registered at her husband's hotel between November 13 and November 22. Mrs. Holcomb concluded her testimony and went home to care for her ill husband.

The next witness the prosecution called was Charles W. Cool of the Cool Insurance Agency in Glens Falls. Cool was well respected locally, having been elected the first mayor of the city of Glens Falls in 1908. He testified regarding insurance policies issued to Wyatt and his wife on their property in Huletts Landing and bank drafts paid to Wyatt for the losses the fire had caused.

His testimony on direct examination could be considered rather bland, as Bascom took him through the important specifics of the insurance policies on the property and the amounts paid out to cover the losses. Cool testified that Wyatt had received $37,000 in total to cover the damage the fire had caused.

A 1908 caricature of Glen Falls mayor Charles W. Cool from the *Morning Post*. Cool owned an insurance agency and was called to testify about Wyatt's fire insurance. *Courtesy of the Cool Insuring Agency, Inc.*

Kellogg began his cross-examination endeavoring to show the witness that, after the insurance companies had paid for the loss, the Wyatts lost about $7,000 when the cost of reproducing the property was calculated. The prosecution objected at this point, arguing that Mr. Cool was not familiar with the figures produced by the defense. These objections were sustained by Judge Rogers, and Cool was not allowed to answer the question.

Kellogg was successful in getting Cool to admit that the hotel was rebuilt in 1916, that hotels were considered hazardous risks in the insurance business and that many hotels had burned on Lake George within the past few years. Mr. Cool also admitted in regards to questions about Wyatt's character that he had always heard the defendant's character spoken about in a favorable manner. Cool also testified further in cross-examination that about $30,000 of the insurance money went to Henry W. Buckell, the former owner, who held a mortgage on the property.

District Attorney Bascom then proceeded to question Cool again on direct examination, soliciting testimony that the newspapers picked up on. Mr. Cool stated that prior to the fire in Huletts Landing, he had never heard that Wyatt was associated with a fire in another one of the hotels Wyatt owned. However, after the fire in Huletts Landing, he had heard that Wyatt had had several fires in his properties and that Wyatt had been fully insured on all of them.

Kellogg was not about to have this be Cool's final comments before the jury. On further cross-examination, he asked Cool if he had reinsured Wyatt when the new hotel in Huletts Landing was rebuilt. Cool responded that he had indeed insured the new hotel built in 1916.

Bascom would not let Cool leave the stand without a follow-up question. Continuing his direct examination, Bascom wanted to know whether the new insurance policies were in the name of William H. Wyatt. Cool stated that the policies had been transferred to Wyatt's son, Arthur Wyatt, within the past few days.

However, it was Kellogg who solicited Cool's final response of the day. Kellogg wanted to know if the policies were written by the same insurance companies that had covered the old hotel. Cool responded that the same insurance companies that had previously written the insurance on the old hotel were willing to carry the insurance on the new hotel.

Cool was dismissed after the blistering back-and-forth had finished. Bascom may have asked the first question, but it was Kellogg who had asked the last.

The next witness called by the prosecution on the trial's first day was Henry W. Buckell, who sold the hotel to Wyatt and his wife in 1913. He testified that he had sold the property to Wyatt for $45,000 and took a mortgage in return as part payment. He stated that at the time of the fire, his interest in the mortgage was $30,000, and he had received this amount out of the insurance money.

On Kellogg's cross-examination, Buckell testified more specifically that Wyatt had paid $5,000 in cash upon purchasing the property and $10,000

Henry W. Buckell (kneeling, far left), the former owner of the Hulett Hotel, was called by the prosecution as a witness in the state's case. *Kapusinski family collection.*

more on the purchase price between 1913 and the time of the fire. There had also been a small note for $1,400 with interest for supplies that Wyatt had purchased with the hotel. Buckell said the hotel always did good business, and Wyatt always met his payments as they became due.

Mr. Buckell said that Wyatt had spent considerable money on repairs and improving the sewage system before the fire. New buildings and sleeping accommodations were constructed, and Mr. Buckell said that the defendant had spent between $4,000 and $5,000 for improvements before the fire. He said that Wyatt had good seasons and frequently had more guests than the property could accommodate. Buckell estimated that the total occupancy was 375 persons when the fire occurred. He continued that the hotel was in better repair at the time of the fire than at any other time he owned the property.

Kellogg was able to garner testimony from Buckell that a chimney in the main hotel had given him great trouble, and about every four years it would burn out and scorch the woodwork in the parlor and attic. He said he thought the heaving of the ground in the winter had opened seams in the chimney and let sparks escape.

Mr. Buckell continued that a pipe hole had been cut in the chimney in the second floor to make it possible to set up a stove. The chimney led to the roof from a fireplace, and twelve to fourteen years ago, the chimney had burned out and burned the woodwork in the attic. He also testified that in

1912, there had been a back draft in the fireplace that started a fire in the parlor and burned the roof and woodwork in the attic. On both occasions, the fire was extinguished. He said that when he sold the property to Wyatt, he had warned him that the chimney was dangerous. Kellogg concluded his cross-examination.

Bascom continued his questioning of Buckell, who, after selling the hotel to Wyatt, had retained other property in Huletts Landing and competed against Wyatt for tourists. When asked by Bascom why he didn't work with Wyatt, Buckell responded that the two could not do business together because they could not agree, and both employed their own respective legal counsels. He stated that the value of Wyatt's land after the fire was between $3,000 and $10,000.

On cross-examination, Buckell stated that the value of the property before the fire, after Wyatt had made repairs, was $50,000. The new hotel, he said, was more cheaply constructed than the previous hotel, and the cost of reconstructing the new hotel had been a very expensive proposition.

With this witness, Bascom would get the last question. He finished by asking Buckell if he had heard any complaints from the tourists about Wyatt's new hotel. Buckell answered that even though the new hotel had had a good season in 1916, he had heard many complaints about poor ventilation and poor construction. Bascom's witness was then excused from the stand.

The last witness of the day was Loren Angell, an accountant in the Albany office of the New York Telephone Company, who was called by the prosecution. She produced ledger sheets showing the telephone accounts of John Sharpe for both his home and livery business in 1915. The sheets showed total charges for toll calls, but the company had destroyed the documents enumerating the calls and showing the dates they were made. In response to questioning from District Attorney Bascom, Angell stated that the original toll tickets showing by whom and to whom the calls were made were destroyed under a rule of the Interstate Commerce commission, but entries for the total number of the calls were entered on the ledger sheets. For the month ending on November 20, Sharpe's toll charges at his residence amounted to $1.69, and at his business the total was $1.90. The card was received into evidence over the objection of the defense.

On cross-examination, Kellogg was able to get Angell to admit that the toll calls appeared on Sharpe's bills for months other than November. After this apparent contradiction, court was adjourned until the following morning. The jury was sent home with instructions by Judge Rogers not to discuss the case with anyone or read about the day's events in the newspaper.

Wednesday Morning, May 16, 1917

At 10:00 a.m., the Hudson Falls courtroom came alive as court came back into session.

Henry W. Buckell, the former owner of the hotel, was recalled by District Attorney Bascom. The district attorney began by asking Mr. Buckell to describe the location of the chimney that, the day before, he had testified had caused him trouble. Buckell also testified that when he last owned the property in 1913, he had carried fire insurance of $25,000 on the buildings and contents.

Bascom endeavored to solicit testimony from Buckell as to the value of the land and buildings adjoining the Wyatt property, but Judge Rogers sustained an objection by the defense. Buckell did state that the old buildings were of the old style and that, outwardly, the new hotel appeared to be more modern.

However, on cross-examination by Kellogg, Buckell stated that the inside of the new hotel was poorly constructed and that the old hotel had been a much better building. He also said that in 1913, he had carried fire insurance of $25,000 on the buildings alone.

Once again, Kellogg had raised some inconsistencies.

The prosecution next called John D. Sharpe, who had been indicted with Wyatt on the arson charge, as one of its principal witnesses.

Sharpe testified that he was a resident of Round Lake, New York, and thirty-three years old. He was operating a livery stable at 144 North Pearl Street in Albany, New York, and he stated that he was married but had not lived with his wife for four years. He said he had known Wyatt for seven or eight years after becoming acquainted with him at Round Lake while engaged in the livery business. He testified that in 1912, he had run a livery for Wyatt at Wyatt's hotel in Lake Bomoseen. After the summer season ended, he returned to Round Lake and continued his business relations with Wyatt.

In the fall of 1914, Sharpe said Wyatt visited him at his home, a rooming house in Albany run by Mrs. Cornelia Gries. Sharpe said he was living with Mrs. Gries at the time Wyatt stopped by. He said Wyatt came alone, and no one else was originally present. However, as they were talking, Mrs. Gries's sister, Mrs. Pregent, came in. All Sharpe recalled from this conversation was that Wyatt told him he was at that time living in Troy, New York.

Sharpe testified that in November 1915, he had called Wyatt in Huletts Landing on the telephone from his home in Albany. Sharpe said he wanted

Keeler's Hotel in Albany, circa 1908. John Sharpe testified that he had discussed plans with Wyatt to burn the Hulett Hotel at the Keeler days before the fire. In an ironic coincidence, Keeler's Hotel burned down in 1919. *U.S. Library of Congress.*

to borrow some money from Wyatt so that he could start a livery business in Albany. Wyatt told him that he would meet him in Albany. According to Sharpe, about a day or two later, Wyatt called him on the phone from Huletts Landing and told him he was coming to Albany that afternoon and asked him to meet him at the Albany train station. This, the witness said, was three or four days before the fire. Sharpe said that he and Mrs. Gries had met Wyatt at the train station, and the three went to the Keeler Hotel for lunch. Mrs. Gries sat in an adjoining room as he and Wyatt spoke.

The witness said that Wyatt told him he was in trouble with the hotel's sewage disposal plant, and unless he could get it fixed, he would have to dispose of the hotel and build in some other place.

Sharpe said Wyatt asked him to help him to get rid of the hotel, but Sharpe refused. He said that Wyatt also mentioned that he wanted the hotel destroyed in the presence of Mrs. Gries and that Mrs. Gries took it as a joke. According to Sharpe, Mrs. Gries replied to Wyatt that Sharpe didn't have enough nerve to do it. Sharpe stated that Wyatt then said he was sorry

November 1915

Sun	Mon	Tue	Wed	Thu	Fri	Sat
	1	2	3	4	5	6
7	8	9	10	11	12	13
14	15	16	17	18	19	20
21	22	23	24	25	26	27
28	29	30				

John Sharpe testified that he had traveled to Huletts Landing twice, once on Friday, November 12, with Cornelia Gries under an assumed name and again on Saturday, November 13, with Wallace Strickler. The fire occurred on November 14.

he had mentioned destroying the hotel and that he felt Sharpe would lose confidence in him. Sharpe stated that Wyatt then told him that Sharpe was the only person he knew whom he could ask to help destroy the hotel. He said that Wyatt did not mention the word "burn" in that conversation.

A few days later, Sharpe testified, Wyatt had asked him to go to Huletts Landing and look around. Sharpe recounted that Wyatt wanted him to see that preparations had been made in one of the rooms to destroy the hotel. He said that Wyatt told him a mattress soaked with oil, together with a quantity of paper and wood chips, had been prepared in one of the rooms. He claimed Wyatt told him that a candle should be lit and allowed to burn down to the paper. Sharpe stated that this conversation with Wyatt took place in Albany and that Wyatt gave him forty dollars for his expenses for the trip. Sharpe stated that Wyatt said nothing about compensation and that at this time he told Wyatt he would not promise to destroy the hotel. This conversation, according to Sharpe, took place on Thursday, November 12, and he went to Huletts Landing the following day, Friday, November 13.

Sharpe's testimony was actually in error, as Thursday was November 11 and Friday was November 12. This error was immediately recognized by

Sharpe, Gries and Strickler, as they traveled by train, would have dressed in contemporary clothing styles, like these train travelers circa 1915. *U.S. Library of Congress.*

Wyatt's defense attorney, Kellogg, who called the court's attention to the fact that November 13 was not Friday. The witness insisted, though, that he went to Huletts Landing on Friday with Mrs. Gries and registered at the Clemons Hotel as James Smith and wife. Sharpe indentified the handwriting on the previously admitted register as his own.

The Whitehall D&H train station, circa 1910. Sharpe and Strickler testified that they took the train to Whitehall and walked to Huletts Landing. *Courtesy of the Whitehall Historical Society/Ray Rose.*

Sharpe then said that he went to Whitehall by train the next day with Wallace Strickler, an employee of his, and the two walked to Huletts Landing, arriving at the Landing about one o'clock in the morning on Sunday. He claimed they went into the hotel through a door, which he said Wyatt had told him would be open. Sharpe said that he had purchased an electric flashlight and candle. The candle, he said, he placed near the oil-soaked mattress and wood chips in the presence of Strickler. After lighting the candle, he placed a newspaper over the flames so it would communicate with the chips and mattress.

Sharpe and Strickler then walked back to Whitehall, getting some breakfast in a restaurant there, and returned to Albany during the day.

The next week, Sharpe said, he met Wyatt in Albany, and Wyatt gave him $200 and told him he would do well by him when he collected the insurance. The witness said that Wyatt had never before told him he would give him money and that he had not expected anything of the kind.

Sharpe said that Wyatt told him before he went to Huletts Landing on Friday that the hotel would have to be destroyed before Monday morning because Arthur Wyatt was to return to finish work on the sewage disposal system. Sharpe also said that Wyatt had arranged to destroy the place himself until Sharpe telephoned about a loan.

The Arlington Hotel in Whitehall was located across the street from the D&H train station. Sharpe and Strickler may have eaten there hours after the fire started. *Courtesy of the Whitehall Historical Society/Ray Rose.*

The witness also corrected himself about the date he was at Clemons, stating that it was Friday, November 12.

When Sharpe saw Wyatt a second time after the fire, Sharpe said Wyatt gave him between $150 and $200. A short time after this, he said he borrowed $125 from Wyatt, giving him a note, which he had not yet repaid.

In March 1916, Sharpe said he borrowed $150 more from Wyatt, giving another note, which also remained unpaid. On Christmas Eve 1915, Wyatt visited him and gave him $100 as a present.

Sharpe said that Wyatt never owed him any money and that there was no arrangement for compensation for setting the fire.

Sharpe said that he had not spoken to anyone about the fire until he was arrested. For a considerable period, and up to about seven weeks earlier, Sharpe said he had worked as a chef for Wyatt at the Hotel Trojan in Troy. Sharpe said he left Wyatt's employ in March 1917 and was arrested on April 25, 1917. The day he was arrested, Sharpe had returned to Round Lake and was working at exercising horses. He said an officer came for him and brought him to Hudson Falls, where he was locked up for twenty-four hours. It was after being arrested and being in jail overnight that he made his statement about what had transpired.

Bascom concluded the prosecution's direct examination of Sharpe by getting him to repeat a remark that Wyatt had supposedly made to him. According to Sharpe, on one occasion after the fire, Wyatt told him that he would prefer to be dead rather than have his family know anything about the fire.

It would now be the defense's turn to cross-examine Sharpe. At exactly 11:10 a.m., after a ten-minute recess, Wyatt's attorney, Kellogg, commenced the cross-examination of Sharpe.

Sharpe admitted that he had left his wife and two infant girls in Round Lake in 1912 or 1913. Two months later, he went to live with Mrs. Gries, with whom he said he had been friendly for a year or more. The scandalous nature of this admission would not have been lost on the conservative Washington County jury. He said he became acquainted with Wyatt in 1911 and had a few business transactions with him between that time and November 1915, but there was nothing about their relations that would cause Wyatt to believe that he would do anything criminal. He said that from March 1913 to November 1915, he saw Wyatt once. That was the time Wyatt called on him in October 1914.

Sharpe stated that the date he called Wyatt on the telephone in Huletts Landing was either November 6 or November 8, 1915, and prior to calling Wyatt on this occasion, he had borrowed money from him on other occasions. Regarding Wyatt's telephone call to him from Huletts Landing on November 10, Sharpe testified, Wyatt did tell him that he had men making repairs to the sewage disposal system at the hotel.

In another triumph for Kellogg's cross-examination, Sharpe said that Wyatt told him he would have to have the hotel burned and asked him to do the job. In this, the witness contradicted his direct testimony in which he had said that Wyatt did not use the word "burn" in the conversation but talked instead of "destroying" the property.

As the cross-examination continued, Sharpe said Wyatt did not agree to pay him any specific amount for the job but after the fire gave him several hundred dollars. The witness admitted that he had obtained part of the money from Wyatt as loans, signing notes to Wyatt in return as security, and that these notes were still unpaid.

Sharpe also appeared to contradict his own testimony again when he said that Wyatt told Mrs. Gries that he wanted to sell the property quick because of his trouble with the sewage disposal system and that he wanted Sharpe to sell it. Sharpe said that Wyatt said nothing to Mrs. Gries about burning the hotel.

At noon, a recess was taken, and court was adjourned until 2:00 p.m. It had been an exciting morning, but now it was time for everyone to go to lunch.

WEDNESDAY AFTERNOON, MAY 16, 1917

The trial resumed at 2:00 p.m. with Judge Rogers allowing the defense to call two witnesses out of the regular scheduled order.

William H. Anderson, one of the owners of the *Troy Times*, testified as a character witness for the defense. Mr. Anderson testified that Wyatt was a person of good character and that he had talked with many persons regarding Mr. Wyatt and had always heard him spoken of as a good citizen.

W.W. Blackmer of Saratoga Springs was also sworn as a character witness for the defendant and testified that Mr. Wyatt had always borne a good character throughout this section of the state.

After these two witnesses, Sharpe retook the stand for further cross-examination by the defense at 2:40 p.m. Kellogg began by wanting to learn more about Sharpe's meeting with Wyatt in Keeler's Hotel in Albany days before the fire. Sharpe stated that he and Wyatt were in a private dining room, separated from adjoining rooms only by latticework, and that Mrs. Gries was in the next room. When the three were leaving, Sharpe said Wyatt arranged to meet him the next day.

The witness said that up until the time he went to Huletts Landing to look at the hotel, he had not been in the vicinity before. Kellogg took Sharpe over the ground covered in his two trips to Huletts. Sharpe denied that he knew that there was a caretaker in the house and denied that he had told his companion, Strickler, that there were people asleep in the house.

Regarding Wyatt's hotel proposition, Sharpe said that all he had told Mrs. Gries was that Wyatt wanted him to sell the hotel.

As the cross-examination continued, Sharpe said that on the morning of the day he went back to Huletts with Strickler, he told Strickler that a man had asked him to burn a building, and Strickler wanted to go along. Sharpe said he had known Strickler about one year. Strickler, he said, wanted to know how much there was in it, and Sharpe said he was unable to tell him. Sharpe stated that Strickler seemed to be "pretty game" and offered to go for $100; he agreed to pay him the $100.

Kellogg questioned Sharpe about whether the moon was out on the night he and Strickler went to Huletts, but the witness's memory on the subject got him considerably confused. Upon entering the hotel, Sharpe said he and Strickler had used a pocket electric light and made no attempt to go quietly.

Sharpe testified further under cross-examination that he was unable to state whether there was a bed in the room where the fire was started. In regards to the lit candle, Sharpe testified that he thought the candle would burn eight or nine hours before it ignited the paper on the floor but denied that he knew how long it would burn from experience. He also stated that he had carried a candlestick in which the candle was placed to burn down to the paper.

Sharpe denied that he had told Strickler that he was going to Huletts to buy horses. He also denied that, when he fell over a wheelbarrow upon leaving the hotel, he told Strickler they couldn't make noise as there were people in the house.

Sharpe admitted that he and Mrs. Gries had a falling out after the fire and that she had told Strickler she would get even with "you damned criminals."

On November 18, 1915, Sharpe said, Wyatt gave him $200 and said he would see him again when he got his insurance. He said Wyatt asked him how he had set the fire alone, and he told Wyatt that he had not done it alone but that Strickler had assisted him. Sharpe stated that within three weeks after the first money was given to him, Wyatt gave him some more money—he believed the amount was about $200—which he said he borrowed from Wyatt and gave notes for.

Sharpe said that he had never been arrested before Constable Dunn arrested him in April. Before his arrest, Sharpe said he had talked with only Strickler about the fire between the time of the fire and his arrest. Strickler, according to Sharpe, talked about the fire every time he had a chance and seemed to feel that Sharpe had gotten him in a lot of trouble. Sharpe continued that Mrs. Gries did not know that he had set the fire, and though she tried to discuss it with him after seeing it in the paper, he refused to talk to her about it.

Sharpe said that District Attorney Bascom had not promised him immunity if he put it off on Wyatt. He said that after his arrest, a deposition—claimed to have been made by Strickler—was read him, accusing him of the crime, and he then told his story to the prosecuting attorney. He stated that he did not know whether he was to be freed for telling the story on the stand. He said that Strickler's deposition contained many things that were not true. He denied again that he had planned to put the blame on Wyatt. Kellogg ended his cross-examination at this time.

Bascom then began a further examination of Sharpe. On redirect examination, Sharpe said that when he finally decided to make his statement to the district attorney, the following people were present in the room: F.F. Stone of Glens Falls, court stenographer Charles Raymond, Constable James Dunn and others. Sharpe said he went before the grand jury as a witness and signed a waiver of immunity. He testified that Mrs. Gries was in the room when he made his statement and that she made a statement also. He stated that no promise of immunity was made by the district attorney.

Bascom then turned his questions toward the money Sharpe said he had received from Wyatt. Sharpe stated that with the first $200 he received from Wyatt, he paid $100 to Strickler, paid some bills and purchased some harnesses and livery supplies. Bascom concluded his redirect examination of Sharpe.

On further cross-examination by Kellogg, Sharpe said he never saw the statement after it was taken by the stenographer and that he did not know whether it was ever written out. Kellogg ended his questioning, and Sharpe's testimony concluded at 4:40 p.m.

Immediately after Sharpe concluded his testimony, Wallace Strickler, Sharpe's companion at Huletts Landing on the day of the fire, was sworn in by the prosecution. He stated that he was thirty-seven years old, an honorably discharged soldier and a livery stable worker. He stated that his principal duties were driving hospital ambulances.

Strickler testified that he worked for Sharpe in November 1915, and on the day Sharpe went to Huletts to set the fire, Sharpe told Strickler he wanted him to accompany him north to get some horses. They took the train to Whitehall, and upon arriving there, Strickler said they started for Clemons. But Strickler said he did not know but what they were going for horses. Arriving at the Hulett Hotel, he said Sharpe took him inside, and while he stood in the hall, he saw Sharpe light the candle. Even then, he claimed he did not realize that Sharpe was "firing" the place. Strickler said he left the hotel ahead of Sharpe, and Sharpe came out a few minutes after he did. Upon returning to Whitehall, Strickler said that he and Sharpe slept in a restaurant near the station for several hours while waiting for a train. About three days after the fire, Strickler said Sharpe paid him $100, and a few minutes later, he showed the money to Harold Lehar and Bob Sharpe, the latter being no relation to the alleged arsonist.

Strickler stated that when Sharpe gave him the $100, Sharpe told him that Wyatt had been down and paid him $200. When asked if he had some

Another view of the Whitehall train station. After leaving Huletts Landing, Sharpe and Strickler testified that they walked back to Whitehall and returned by train to Albany. *Courtesy of the Whitehall Historical Society/Ray Rose.*

trouble with Mrs. Gries, Strickler stated he had, and he thought Mrs. Gries did not even like herself.

Following the fire, Strickler said he saw Wyatt and his son-in-law visit the Sharpe stable to get a pair of horses. The witness said he and Sharpe parted as friends when Sharpe closed out business in the spring of 1916. Bascom thus finished his direct examination of Strickler.

On Kellogg's cross-examination, Strickler said that Sharpe and Mrs. Gries had trouble almost constantly. He said that a short time after the fire, Mrs. Gries called the stable, and he answered the phone. She threatened to "fix the both of you criminals." Strickler told Sharpe, and Sharpe said, "You don't know her." Strickler stated that later Sharpe told him that Mrs. Gries had "burned a couple of babies or something like that" and he would get even with her.

Strickler denied that Sharpe told him he was going to burn a building and also said that Sharpe made him no proposition to assist him. Strickler continued his testimony by stating that while on the trip to Huletts Landing, Sharpe did not tell him where they were going. The witness said the walk from Whitehall to the place on the lake, which Strickler could not identify by name, was a lonesome one, and he did not blame Sharpe for wanting company.

Kellogg sought to solicit more information about what happened when the fire was actually set. Strickler said that while he was waiting in the hall for Sharpe (who was in the room), he heard a door crank and, thinking someone was approaching, called to Sharpe in the room. He said he believed Sharpe might know the people in the place—but there was no response.

Before going into the room where the candle was lit, Strickler said Sharpe went into another room, but he didn't know why.

After seeing Sharpe light the candle, Strickler said he realized Sharpe was committing a crime but made no protest. However, he did tell F.F. Stone, an attorney from Glens Falls, about the incident a couple of weeks later. Strickler left the stand at 5:20 p.m. without Kellogg finishing his cross-examination. This admission would be important, and Kellogg would revisit it. However, another witness, who could not return the next day, had to be sworn in.

Dr. Martin E. Sargent of Putnam, who in 1915 was the health officer for the town of Dresden, in which Huletts Landing is located, was sworn in next for the prosecution. He testified that in the summer of 1915, he had inspected the sewage disposal system and found that it was in bad shape; conditions were very unsanitary. Complaints had been made by cottagers because of the overflow of the cesspool. Dr. Sargent said that the town health board had adopted a resolution directing Wyatt to make repairs.

Town clerk Ernest L. Clemons of Dresden testified next that Wyatt was ordered to make the repairs before he opened for the 1916 season. After his testimony concluded, court adjourned for the day.

Thursday Morning, May 17, 1917

C.A. Holmquist of Albany, assistant sanitary engineer of the state department of health, was sworn in at the opening of the trial on the morning of May 17. He was called as a witness for the prosecution and described the condition of the sewage disposal system of Wyatt's hotel in 1915 before the fire.

On cross-examination by Kellogg, the witness stated that it would have cost Wyatt only about $400 to install the system recommended by the state department of health. He also admitted that it would have only been necessary to operate the gasoline-powered pump for about three hours per day. This testimony blasted the impression that had been conveyed the previous day by the testimony of Sharpe and Strickler that Wyatt wanted the hotel burned because of trouble with the sewage disposal plant.

Arthur Derby of Benson, Vermont, who in 1915 worked for Charles Holcomb of Clemons, was the next witness called by the prosecution. Mr. Derby was a coachman, a man whose business it was to drive a horse-drawn carriage. Mr. Derby testified about driving Sharpe and Mrs. Gries in a carriage to Huletts Landing a few days before the fire and identified them in the courtroom as the parties he had driven on the day in question.

However, on cross-examination, Derby said that the man who was with the woman he took over to Huletts wore a beard, and they arrived on the train from the north instead of the south, as Sharpe had testified. When they left, they went north, not south, as Sharpe had stated.

As the cross-examination continued, though, Derby admitted he had become confused and that Sharpe and Mrs. Gries had arrived on the train as Sharpe testified.

On further cross-examination, Derby stated that he had taken three men to Huletts the same week as Sharpe and Mrs. Gries, and these three men came from the north and left on a northbound train.

According to the account of the *Glens Falls Times and Messenger*, the next witness called gave some of the most damaging testimony of the trial. Robert J. (Bob) Sharpe, who in November 1915 was employed by John D. Sharpe in Albany but was in no way related, was sworn in next for the prosecution.

The witness testified that the Saturday before the fire, Sharpe and Strickler had gone away and did not return until Sunday evening, thus apparently corroborating Sharpe's and Strickler's testimonies. Some time after the fire, Bob said, he was at Sharpe's house fixing the furnace, and while he was there, Wyatt stopped by. He heard Wyatt say, "My God, Jack, how did you do it?" He stated that Sharpe replied, "Not the way you told me to and not the way you wanted me to." A few minutes later, the witness said he saw Wyatt hand Sharpe some money. The witness also swore that Strickler, some time after, exhibited $100 to him.

When cross-examined by Wyatt's counsel, the witness said that a quarrel arose sometime later between Sharpe and Mrs. Gries. He stated that when he went into the livery stable at this time, he found Strickler pacing up and down the floor, and Strickler told him about the threats made by Mrs. Gries. He said that he then told Strickler of having seen Wyatt at Sharpe's home. The defense rested its cross-examination of Bob Sharpe. However, the prosecution seemed to have scored important corroborating testimony from a disinterested party.

Arthur Wyatt, son of William H. Wyatt, was called as a witness by both the prosecution and the defense. *Kapusinski family collection.*

The prosecution continued its case by next calling Arthur Wyatt, the son of the defendant. Arthur Wyatt testified that the week preceding the fire, he was in Troy. The witness testified that the hotel had closed in the middle of September and that a Mr. and Mrs. Foster remained in the hotel as caretakers, having quarters on the first floor. He stated that there was a telephone at the hotel in Huletts. He also said that he had known Sharpe for some time before the fire.

When cross-examined by Kellogg for the defense, the witness stated that he had instructed the caretakers to sleep in the hotel every night. He testified that he had a lease for the premises, with an option to purchase, from his father. He stated that he had operated the hotel for the 1915 season for a rental (to his father) of $6,000 and also realized an additional $5,000 profit for himself. The new hotel, he said, he rented in 1916 from his father for $5,500, and he realized a profit of $3,300. The defense succeeded in bringing up an important point: both Wyatts made less money on the new hotel.

Regarding the sewage disposal system, Arthur Wyatt said he and his father had talked for two years about improving the system and were working on it in 1915 when the state health department took up the matter. He stated that work on the sewage system was in progress when the hotel was destroyed.

Concerning the new hotel, the witness said that the structure cost $35,000 to build, and with the furnishings included, the total came to $50,000. On the $30,000 mortgage held by Henry W. Buckell on the old hotel, Arthur Wyatt said his father paid 5 percent interest. In regards to the new hotel, the witness said his father was paying 6 percent on the money he had borrowed to construct it. The younger Wyatt also testified that the new hotel was smaller than the old one and more poorly constructed.

In the fall of 1915, Arthur Wyatt said, his father's health was very poor; he was suffering from heart trouble.

Leases for the hotel property were produced by the defense, and it was shown that in September 1915, the witness had leased the old hotel for the next year (1916), agreeing to pay $6,000 for the year, but when the new hotel was completed, the rent was reduced to $5,500.

When further examined by Kellogg, the witness said that his father had spent more than $7,000 on repairs and alterations and that at the close of the season in 1915, the hotel and buildings were in good repair.

Arthur Wyatt finished his testimony with no further questions from District Attorney Bascom.

The final witness of the morning called by the prosecution was Mac Potter, teller in the First National Bank of Glens Falls. He was sworn in as a witness for the prosecution. He testified that Wyatt gave two notes to Sharpe in 1916; they were discounted by the bank and charged to Wyatt's account.

Thursday Afternoon, May 17, 1917

After a recess was taken for lunch, Kellogg was allowed to continue his cross-examination of Wallace Strickler, which he had not been able to conclude on Wednesday afternoon.

Kellogg started by exposing a possible conflict of interest regarding Strickler's lawyer. While the newspaper accounts are unclear, the attorney Strickler sought advice from, F.F. Stone, also worked for the insurance company that had paid Wyatt for the loss. Kellogg immediately brought this issue to the forefront, questioning Strickler about his knowledge of this and wanting to know if F.F. Stone had endeavored to have the insurance money returned based on Strickler's conversations with him.

Strickler said he did not know what Stone's interest in the case might be and stated that he first told Stone about the fire to obtain advice about how to protect himself. He swore that Stone told him that he would look into the matter. Kellogg ceased his questioning here, most likely because he had successfully introduced a significant element of doubt to the jury about Strickler's motive and credibility. Was Strickler a lackey of the insurance company, doing their bidding to get the insurance money back? On the other hand, if Strickler was credible, why hadn't the attorney started proceedings to retrieve the money? Kellogg had successfully raised the issue. Strickler was dismissed.

Miss Cora Phillips of Huletts Landing was the next witness sworn in by the prosecution. She testified that she was familiar with the old hotel property. She said the fire started about ten o'clock on Sunday morning, November 14, 1915, and she first saw the flames as they broke through the side of the roof, six to eight feet from the chimney. According to the witness, the entire structure was in ruins by noon.

Miss Phillips said she had seen Wyatt at the hotel either Thursday or Friday before the fire standing on the hotel's steps. According to Miss Phillips, the fire did not break out near the chimney that Buckell said was defective.

On cross-examination, Miss Phillips denied that she had had trouble with Wyatt because he refused to let her run the post office in Huletts. However, she admitted that she could not say at what point the fire broke out on the other side of the roof from the direction from which she first saw it.

Kellogg was again successful in planting doubt. Perhaps the witness was not as impartial as she first seemed, and maybe the fire had started near the chimney opposite of where she saw it.

Kellogg was zealous. Bit by bit, he chipped away at the prosecution's witnesses.

Mrs. Adelaide Cooper, the next witness for the prosecution, corroborated Miss Phillips's testimony as to the location in the hotel where the fire broke out. She also swore that she had seen the defendant at the hotel the Friday morning before the fire.

William Dunn of Ticonderoga, a carpenter by trade, was the next witness. He testified that in November 1915, he was at work at Huletts, and during the week before the fire, he saw Wyatt at the hotel. According to his best recollection, he testified that Wyatt was at the hotel either Thursday or Friday before the fire.

Cornelia Gries was next called to the stand by the prosecution. Her testimony would be critically important in corroborating Sharpe's account.

She started slowly, testifying that she was born in 1880, and her home was now in Schenectady, but she had lived with Sharpe in Albany in November 1915. She then corroborated Sharpe's accounts of meeting Wyatt in Albany, the conversation at the Keeler's Hotel restaurant and the trip to Clemons the Friday before the fire. She also stated that the next day (Saturday), Sharpe went away and stayed away until Sunday afternoon or evening.

On Christmas Eve 1915, Gries said that Wyatt called at the home and gave her a fifty-dollar bill, but she said she did not see Wyatt give Sharpe anything.

The summer of 1916, Gries said she and Sharpe spent the weekends at the Columbian Hotel in Saratoga Springs most every week and always saw Wyatt. During the winter of 1916–17, Mrs. Gries said she saw Wyatt quite often in Troy. She also corroborated Sharpe's testimony about receiving a telephone call from Wyatt the Wednesday before the fire.

Having elicited this important corroborating testimony, Bascom ended his direct questioning of Gries.

Kellogg began his cross-examination of Gries by bringing up the most salacious details of her relationship with Sharpe first. The witness admitted that she was living as Mrs. Sharpe and stated on the record that her name was neither Sharpe nor Gries. This admission would have immediately impugned her moral character and truthfulness in regards to the standards of 1917. Was she Sharpe's common-law wife? Was she married to another man yet living with Sharpe? Had she failed to obtain a divorce? Was she simply a person of low moral character? She had allowed herself to be called Mrs. Gries throughout the trial but now admitted that this was not her name. Kellogg ended this line of questioning here, having successfully raised these issues in the minds of the conservative jurors.

Gries said that before the meeting in Keeler's Hotel, she knew that Sharpe was going to ask Wyatt for a loan.

Regarding having assumed the name of Smith when she and Sharpe went to Clemons and Huletts the Friday before the fire, she said she knew of no reason why they should assume the name of Smith.

She also admitted that she had visited District Attorney Bascom's office on a number of occasions before the trial.

Kellogg seized on this admission and questioned Gries about whether she had been in the office of the district attorney the previous Thursday evening at 11:00 p.m. going over her statement of what she was to testify. Gries denied that she was examining her statement on that occasion and was unable to say how late she was in the office. Bascom jumped to his feet and objected to the implication that he had been drilling the witness on her testimony.

After a short conference with the judge, in which Rogers most likely reminded both sides to remain civil, Gries was dismissed from the stand.

Miss Mary Fagan, a telephone operator who was employed in the exchange in Whitehall in November 1915, was next sworn in and produced a toll ticket for a call from Wyatt's hotel at Clemons to John D. Sharpe in Albany, made on November 10, 1915.

On cross-examination, though, the witness said that there was nothing to indicate who had sent the call from the hotel.

Arthur J. Williams, cashier of the Granville Telephone Company, Miss Fagan's employer, corroborated her testimony regarding the records of the company in relation to the toll call.

Henry W. Buckell was next called to the stand for further cross-examination by Joseph Kellogg about the contour of the old hotel. In many respects, the witness contradicted Sharpe and Strickler regarding the course they said they had taken to get into the hotel. Buckell also contradicted Sharpe about there being a gateway in the pasture through which the hotel could be reached. Sharpe had testified that Wyatt told him he could get in through such a gate.

Buckell said that the Meadow Point Hotel, near the Hulett House, was constructed similar to Wyatt's hotel. It was his opinion that the construction of the hotel described by Sharpe and Strickler as the place they had entered and the course they followed sounded almost identical to the Meadow Point Hotel, not the Hulett House.

District Attorney Bascom had one final chance with his witness. When questioned by the district attorney, Buckell denied that he had been advised that if Wyatt were convicted, he would have to return the $30,000 that Wyatt had paid him from the proceeds of the insurance money to discharge the mortgage.

The prosecution thus rested its case on Thursday, May 17, 1917, at 4:30 p.m. The state had based much of its rationale on the testimonies of John Sharpe, Wallace Strickler, Cornelia Gries and Bob Sharpe. John Sharpe alleged that he had been paid to set the fire by Wyatt, who was intent on collecting the insurance money. The inadequate sewage disposal system for the structure was considered to be the primary motive. The other witnesses seemed to substantiate and give credence to Sharpe's account. Numerous other corroborating witnesses offered testimony, which, in total, appeared convincing. Now, the defense would have its turn.

Before Kellogg began his opening statement, he made two motions to Judge Rogers. The first motion made by the defense was to strike out the ledger accounts for Sharpe's telephone in Albany, produced by Loren Angell on the first day of the trial. Angell had already admitted that the toll calls appeared on Sharpe's bills for months other than November, thus casting serious doubt on their relevance and accuracy. Judge Rogers granted this first defense motion.

Kellogg's second motion was to dismiss the indictment in its entirety, made on the argument that the prosecution had not proven its case. Judge Rogers denied this second motion.

Joseph A. Kellogg, by all accounts, was an extremely gifted orator. In his opening for the defense, he made a scathing attack on Sharpe, Strickler and Gries and ridiculed the idea of Wyatt putting himself at the mercy of "these stable bunch of fellows." He mocked the idea that Wyatt would have the hotel burned when it was doing a most successful business and when he would suffer a huge loss by its very destruction. He asserted that the defense would disprove any motive Wyatt might have had for wanting the property destroyed. Kellogg's opening was brief and to the point.

William A. Lennox of Albany, manager of an office of the General Adjustment Bureau, was the first witness called by the defense. He testified that the insurance companies concluded that the loss on the main hotel property exceeded the amount of insurance carried. On the main hotel, Lennox testified that Wyatt had lost $8,000, according to the computations. On the furniture, the witness stated that the computations showed a loss of about $900. On the other buildings that burned, there were also losses over and above the amount of insurance. There was no cross-examination by the district attorney.

Captain Walter G. Watts of Silver Bay, a civil engineer, testified for the defense that in September 1915, he was engaged by Wyatt to make plans for a sewage disposal plant for the old hotel and that the work was in progress at the time of the fire. Later, he said, he continued the work for the new hotel. He said the cost of doing the work for the old hotel would have been just over $400, while the cost of doing the work for the new hotel was about $600. The plans, he said, met the requirements of the state department of health, and Watts asserted that he had advised Wyatt that it was perfectly feasible to dispose of the sewage as required by the state health department.

Captain Watts testified that he was at the hotel on Monday, November 8, and Thursday, November 11, 1915. He stated that his records showed that Wyatt was at the hotel on November 3, but his diary did not show that Wyatt was there on November 11.

On cross-examination by Bascom, Watts did admit that another system, in addition to his system, had to be installed so that the overall sewage disposal system would work correctly when it came online. He could not state what the cost of that additional system was.

On redirect examination by Kellogg, though, Watts said that the additional system had proved useless, and his system was being used exclusively. Captain Watts was dismissed from the stand. A long day was over.

Friday Morning, May 18, 1917

The day began fair and warm with moderate northeast winds. The defense started the morning by calling Harlan Foote of Huletts Landing as the first witness of the day. He testified that in November 1915, he had been employed by William Wyatt making repairs to the hotel. Foote stated that excavations had been made at the time of the fire for a new sewage disposal system and that lumber for the forms arrived a day or two after the fire.

Willis Foster of Huletts Landing, the caretaker of the hotel in November 1915, was next sworn in as a witness for the defense. He testified that he and his wife slept in the hotel and that they were sleeping in the hotel the night before the fire. He said there was a dog and three puppies in the building with them that night. He stated that he and his wife slept near the front part of the main building. Foster stated that once a day he went through the various rooms in the buildings in accordance with the instructions of Arthur H. Wyatt, William Wyatt's son. The day before the fire, Foster said he went through the building and observed no windows or doors open. He said the door that Sharpe and Strickler testified was open when they claimed to have set the fire early Sunday morning was locked when he inspected it on Saturday.

According to Foster, no noise was heard in the hotel on Saturday night or Sunday morning, and the dogs made no disturbance. The witness stated that there was a wood fire kept in a stove during November and that the stovepipe went into the chimney that Henry Buckell testified was defective.

When the fire broke out between nine and ten o'clock on Sunday morning, Foster said the blaze broke out near the ridge of the roof at a point near the defective chimney.

After a thorough cross-examination by the district attorney, Foster's testimony remained unchanged.

Mrs. Marilla Foster, wife of Willis Foster, was next called to the stand by the defense and corroborated her husband's testimony, stating that the dogs were kept by the Wyatts as watchdogs and were free to roam the hotel. She stated that there were fires in two stoves in the hotel on the morning that the fire destroyed it and that she had used the kitchen stove for baking that morning.

On cross-examination by Bascom, Mrs. Foster stated that she did not believe it would have been possible for Sharpe and Strickler to enter in the hotel without waking her and her husband.

The Fosters would thus prove to be unshakable witnesses for Wyatt. Bascom could not get them to waiver in any regard.

Adelbert Buckell, who was with Willis Foster when he discovered the fire, was the next witness called by the defense. He swore that the fire broke through the roof near the defective chimney. The witness said he recalled the chimney burning out when his brother, Henry W. Buckell, had the property, and smoke came through the cracks in the chimney in the attic. His brief testimony concluded without cross-examination.

George Griffiths of Albany, a waiter at Keeler's Hotel, was called next by the defense. He testified that Mrs. Gries could not have made use of the rooms she claimed to have used when she said Sharpe and Wyatt were there. Griffiths said that the rooms in question were used by ladies with escorts, and two men were never allowed to occupy the rooms in question together. Once again, there was no cross-examination, Bascom was saving his energy for the next witness.

The defense of any case is purely a device of the attorney of the accused. The law casts no burden on the defendant to prove anything. The entire burden of proving the guilt of a person is with the prosecution. In many instances, the defense, believing that the prosecution has failed in its burden, will simply not present a case. Lawyers and judges understand that the accused has every right to remain silent; indeed, it is a protection the Constitution provides to all. One wonders, though, do people not trained in the law really appreciate this? Many people think, "If he is innocent, why doesn't he just take the stand and say so? What has he got to hide?" The defense in this case did not have this option. The prosecution's case against Wyatt appeared strong. William H. Wyatt would take the stand to tell his side.

Wyatt was sworn in on his own behalf at 10:35 a.m. The *Post Star* reported that at first he appeared "slightly nervous." In response to questions from Kellogg, he testified that he was fifty-six years old and was born in the town of Cambridge, Washington County. He stated that he had known John D. Sharpe since 1912 in a business way, when Sharpe ran a livery for him at Lake Bomoseen. Prior to that time, he had hired Sharpe to take him out in carriages as he made real estate transactions at Round Lake.

At Lake Bomoseen, Wyatt said he had had trouble with Sharpe toward the end of the season because Sharpe had taken liquor from the hotel bar and was familiar with one of the girls employed at the hotel. However, Wyatt said no ill feeling existed when Sharpe left.

Between 1912 and November 1915, Wyatt said it was about two years before he had any further business relations with Sharpe. He said Sharpe called on him at the Clark House in Troy and told him he was running a stable in Albany. Wyatt said that he called at Sharpe's stable once after that and before the fire to look at a pair of horses, which he bought after the fire.

In November 1915, the witness said he went to Huletts to vote, as did his son, and they were accompanied by their wives. He stated he was suffering from heart trouble, and his health was poor. While at the lake, Wyatt said, Captain Watts met them and went over plans for the new sewage disposal plant.

On November 8, 1915, Wyatt said he and Mrs. Wyatt returned to Huletts while he supervised repairs that were being made, and he also ordered lumber for the forms. While there, Wyatt said he got a telephone call from Sharpe, and Sharpe told him he wanted to see him on business. Wyatt told him he could not tell just when he would leave Huletts but would call him later. Wyatt testified that he called Sharpe a day or two later and told him that he could see him at the Columbian Hotel in Saratoga Springs any day during the week. He stated that Sharpe met him at the Columbian Hotel during the week before the fire and asked him to leave him some money. Wyatt told him, according to his testimony, that it had been a big expense making repairs to his hotel property and that he could not accommodate the request at that time but might be able to do so later.

Wyatt swore that he and Mrs. Wyatt, his son and his wife went from Saratoga to Troy and were there when he received news of the fire.

Wyatt denied that he had hired anyone to set the fire and asserted that he had suffered a heavy loss as a result of it. He said he had expended $15,000 on the property and another $12,000 for improvements and furnishings. He denied that he was in Albany the week before the fire or any other time to meet Sharpe and Mrs. Gries.

Wyatt said that the question of the sewage disposal plant was not worrying him because Captain Watts had told him the system would cost only about $800.

After the fire, Wyatt said he saw Sharpe in Albany, and Sharpe wanted some money. According to Wyatt, he loaned Sharpe some money by check, took Sharpe's note, discounted it and put it in the bank. Wyatt said Sharpe later borrowed more money and gave another note. Wyatt said that small payments had been made by Sharpe from time to time, as shown by the note given the bank in renewing the old notes.

Wyatt denied that he had ever given Sharpe any other money as a Christmas present or for any other reason. He denied the gift testified to by Mrs. Gries and said he had never visited Sharpe's home.

According to Wyatt, on Christmas Eve 1915, he was at his home and at the home of his son. He directly contradicted the testimony of Sharpe and Gries by saying that he was not in Albany that day but was instead out shopping with members of his family.

In October 1916, Wyatt said he took possession of the Trojan Hotel in Troy. He stated that Sharpe worked for him from December 1916 until March 13, 1917, and that Sharpe was paid ten dollars a week and received half his room.

Wyatt said he had trouble with Sharpe again because Sharpe frequented the barroom so much. He said he noticed that Sharpe had been acting sullen for some time, and at one time Sharpe said he thought his wife (Mrs. Gries) was crazy and that he would have to take her away. A few days later, Sharpe told Wyatt he was thinking about leaving. Wyatt said that he had always believed Mrs. Gries was Mrs. Sharpe until he was told differently earlier that spring.

He stated that plans for rebuilding the hotel were commenced immediately after the old one burned down. He stated that he had run the hotel himself in 1913 and 1914 and realized $10,000 gross profit, out of which interest, repairs and expenses had to be met.

To construct the new hotel, Wyatt said he had trouble getting money. The First National Bank of Glens Falls turned him down three times until he got in touch with the president, Byron Lapham. According to Wyatt, Lapham called a meeting of the directors of the bank, and they decided to let him have part of the money if the Finch-Pruyn Company would loan him the rest, which Finch-Pruyn ultimately did. Wyatt said he gave a mortgage of $30,000 as security, and the new hotel was ready for the 1916 season. He stated that he rented it to his son, Arthur, for $5,500 a year.

Wyatt was excused from the stand as another witness who couldn't remain at the trial was sworn in.

Attorney Jarvis O'Brien of Troy, formerly of Fort Edward, counsel for the Boston and Maine Railroad in New York State, was sworn in as a character witness for the defense. O'Brien testified that he was fifty-two and had been assistant district attorney and then district attorney in Rensselaer County for three terms. He stated that he had known Wyatt for twelve to fifteen years. O'Brien testified that he had always heard Wyatt spoken of as a good and honest citizen and stated that the defendant had a splendid reputation for honest dealing.

Wyatt again took the stand for questioning by Kellogg. He claimed that when he was arrested for the crime, he was locked up for three days and

Jarvis P. O'Brien, the former district attorney for Rensselaer County, was called as a character witness by the defense. *The Argus Company/Google.*

three nights. The first night, he said District Attorney Bascom told him he would have to furnish $30,000 or $40,000 bail, but Supreme Court Justice Salisbury released him on $10,000 bail.

At 11:30 a.m., District Attorney Bascom commenced his cross-examination of Wyatt with questions about the Glenwood Hotel on Lake Bomoseen. Wyatt stated that in the fall of 1912, his hotel at Lake Bomoseen had indeed burned down. He vigorously denied that the night before the Glenwood fire, the water in the hotel had been turned off. Concerning other fires at his hotels, Wyatt said that he went into business in Troy in 1901, and there were a couple of slight fires in his establishments, but none of any consequence.

He admitted to having a small fire in a restaurant he owned on King Street in Troy and that a farmhouse he owned in Round Lake had burned down.

In regards to Sharpe's testimony that Wyatt came to a rooming house in Albany and met with him, Wyatt stated that he knew Molly Pregent, who worked for him at Lake Bomoseen, and admitted that when he was in Albany to look at a pair of Sharpe's horses, he met Molly, her "boy" and Mrs. Gries and took them to Albany's Empire Theatre. However, he denied that he spent the night at Sharpe's home.

Wyatt was quite clear that whenever Sharpe and Mrs. Gries were at the Columbian Hotel in Saratoga during 1916, they always registered and paid for their lodging and meals just as any other guests.

At noon, a recess was declared until 1:30 p.m., with further cross-examination of Wyatt by Bascom still to come.

Friday Afternoon, May 18, 1917

At 1:30 p.m., Bascom resumed his cross-examination of Wyatt by exploring Wyatt's personal net worth. Much of Bascom's afternoon time was taken up by William Sears, who was assisting Bascom, digging through boxes of records.

Wyatt said that he was paying 6 percent on the $30,000 mortgage held by Buckell, interest on a loan of about $5,000 to the First National Bank of Glens Falls and interest on a mortgage of $4,700 on the Hall House property in Whitehall, which he had recently purchased before the trial started.

He stated that when Sharpe wanted a loan, Sharpe asked to borrow $200, but because Wyatt did not feel he could spare this amount, he had told Sharpe he could not let him have it.

During the summer of 1915, before the fire and after repairs had been made and furnishings updated in the amount of $12,000, Wyatt said he added about $9,000 in fire insurance.

The district attorney concluded his questioning of Wyatt at 2:40 p.m., and Kellogg only wanted one short follow-up question.

Kellogg finally raised the issue of the Glenwood Hotel fire on Lake Bomoseen. Specifically, he wanted to know what the economic impact of that fire was on Wyatt. Wyatt answered that he had lost about $10,000 on the Glenwood Hotel fire. Wyatt left the stand a minute or two later. His testimony was concluded.

Wyatt's daughter, Mrs. Granville, testified next for the defense. She said that she was at the Columbian Hotel in Saratoga when her father, mother, brother and sister-in-law all returned from Huletts Landing a few days before the fire. Bascom did not ask her any questions.

Wyatt's wife, Ella Haynes Wyatt, was sworn in next. She stated that she and her husband had been married in Poultney, Vermont, thirty-seven years earlier and corroborated her husband's testimony in all regards concerning his visits to Huletts in November 1915.

Mrs. Wyatt testified that on Christmas Eve 1916, her husband was at home in Troy and could not have been in Albany as Cornelia Gries and John Sharpe alleged.

Mrs. Wyatt could not have been a more powerful witness in defense of her husband. Bascom, once again, did not ask her any questions.

Wyatt's son, Arthur Wyatt, took the stand next and corroborated the testimony given by both his father and mother. Newspaper accounts make no mention of the district attorney making any cross-examination of the witness.

Mary Wyatt, Arthur Wyatt's wife and William Wyatt's daughter-in-law, was called next to the stand by the defense. She testified that she had herself padlocked the door that Sharpe and Strickler claimed to have found open at the close of the 1915 season and carried the key herself.

Finally, Bascom asked a question of a Wyatt family member. Mary Wyatt admitted under cross-examination by the district attorney that she last saw the padlocked door in question eleven days before the fire.

Howard Slocum of Easton was sworn in next and testified that he had known Wyatt for forty years and always knew him as a man of good character. Slocum swore that he had always heard Wyatt spoken of as an honest, fair-dealing person.

Arthur W. Sherman, cashier of the First National Bank of Glens Falls, was the next witness. He testified that it was part of his duties to ascertain the character of customers, and the bank had loaned Wyatt money with and without security or collateral. Sherman stated that when Wyatt's account was solicited by the bank, he had made the usual inquiries to determine Wyatt's reputation and found the defendant's reputation for honesty and fair dealings to be good.

On cross-examination by Bascom, the witness stated that Wyatt owed the bank $19,000 and was amply secured with plenty of margin. He said the bank was not worried about Wyatt's account.

Sherman was Kellogg's last witness. He rested Wyatt's case at 4:20 p.m.

The district attorney was not yet finished, however. John D. Sharpe was re-called for the prosecution in rebuttal to one point. Sharpe denied that he went to Saratoga and met Wyatt at the Columbian Hotel on November 10, 1915. There was no questioning by the defense.

Both sides rested their cases and agreed to wait until the next morning to sum up their respective sides. The defense renewed its motion to dismiss the case once again. The judge again denied the motion. The day concluded with Judge Rogers warning the jurors not to discuss the case with anyone, and court was adjourned until nine thirty the following morning.

Saturday Morning, May 19, 1917

As the participants returned to the courtroom for the conclusion of the trial on Saturday morning, the ground was wet from rain the night before. The headline of the Saturday newspaper read, "President Signs Army Draft Bill, Registration Scheduled for June 5th, All Males Between the Ages of 21 and 30 Years Inclusive, Must Enroll for Military Service. General Pershing Going to France."

The crowd seated in the courtroom was even larger than usual in anticipation of the end of the trial and the possibility that a verdict could be reached soon. Court began promptly at 9:30 a.m. with both sides' summations scheduled. Wyatt seemed calm but sat upright in his chair. His family sat in the first row behind him.

Joseph A. Kellogg was a talented speaker who was also twenty years older than Bascom. He had extensive trial experience and knew how to give a powerful summation. He had tried many cases and was well paid. An account of his conclusion in the *Glens Falls Post Star* states that he gave it with "tremendous force," which got the "attention of the jury."

Kellogg started by attacking the credibility of the three main witnesses. Sharpe was nothing more than a disgruntled former employee who had been dismissed for being loose with the bar liquor and hotel staff. He had abandoned his wife and children and had taken up with a woman who couldn't even testify as to what her own name was.

Strickler was a person who would say anything that his employer, Sharpe, told him to say and had so little credibility that when he told his outlandish story to his own lawyer, who also happened to work for an insurance company that had paid Wyatt, the lawyer did not start any proceedings to have the money returned.

Cornelia Gries received some of the worst scorn. Kellogg reminded the jury that Strickler had testified that Gries had "burned a couple of babies or something like that" and had made threats against both Sharpe and Strickler. Worse yet, she had run off with a married man who had left his own children to take up with her. Her story about overhearing conversations that Wyatt allegedly had with Sharpe didn't hold up because the Keeler's Hotel staff had testified that the room was not arranged for meetings like the one she testified to.

Kellogg placed special emphasis on Sharpe's motive to "fire" the hotel. According to Kellogg, it was Wyatt's refusal to hand Sharpe money that had caused Sharpe to burn the structure or fabricate his tale. The hotel was a profitable enterprise that was producing a reasonable return for both Wyatt and his son.

Yes, Wyatt had loaned Sharpe some money, but it was a well-documented loan that Sharpe would have to repay. Wyatt was simply trying to help someone from a lower status in life get a leg up. This wasn't the work of a criminal mind. Wyatt was a good citizen. Many upstanding people had testified to Wyatt's impeccable reputation.

Notwithstanding the unsavory cast of characters that had tried to ruin Wyatt's good name, Wyatt had no motive to destroy a successful business, and there was no clear proof that arson had even occurred. Testimony had been given that showed that the fire resulted in a net loss for Wyatt. He was the true victim. Yes, he had the misfortune of two hotel fires in a short period, but Wyatt was not a criminal. He had been in the act of improving the sewage system when the fire occurred. The former owner had even testified that there was trouble with the chimney, and the neighbors had observed that this was where the fire started.

Kellogg's oratory and trial experience held the jury captive. They could not sentence this good man to twenty-five years' hard labor. He had been through enough. He was a man who invested in communities. His hotels brought tourists into the area. Washington County needed more people like Wyatt. The evidence could only yield one result: Wyatt was innocent and had to be found not guilty.

District Attorney Wyman S. Bascom went next and summarized the prosecution's case. By all accounts, he had the jury's complete attention. Bascom's oral arguments were well thought out and presented with great emotion.

Bascom started by reminding the jury that testimony had been delivered showing that Wyatt had operated the Glenwood Hotel on Lake Bomoseen while acquainted with John Sharpe. The two had gone on to do business together, with Sharpe running a livery at the Glenwood. While Bascom didn't say so directly, the inference was clear. The Glenwood had also been destroyed by fire while Wyatt and Sharpe were there.

According to the district attorney, when Sharpe wrote to Wyatt in November 1915 to borrow some money, Wyatt hatched his plot. Wyatt went to Albany to see Sharpe and made Sharpe an offer to burn the hotel. Sharpe brought along the woman with whom he was living, Cornelia Gries, who overheard part of the details.

Shortly after their meeting in Albany, Sharpe and Gries went to Clemons and registered at a Clemons hotel as Mr. and Mrs. James Smith. They then

Union Station in Albany, circa 1904. The prosecution claimed that Sharpe and Gries returned from Clemons to the Albany train station. *U.S. Library of Congress.*

proceeded to hire a driver and carriage to take them over the mountain to Huletts Landing, where they inspected the hotel. They saw the area and surrounding cottages and then returned to Clemons. They left for Albany by train later that day.

According to Bascom, multiple witnesses corroborated Sharpe's account that on the next day, accompanied by his employee, Wallace H. Strickler, they took the train to Whitehall and walked to Huletts Landing. They arrived early in the morning on Sunday, November 14, 1915, and entered the hotel through an open door that had been arranged with Wyatt in advance.

Bascom stated that they walked to the second floor of the hotel and entered a specific room where they found a mattress soaked in oil. The mattress was partially on the floor and partially leaning against a wall. At one end of the mattress, there was a hole in the floor where there was a quantity of wood chips and paper. Sharpe arranged the paper to lead to the mattress. He then took a candle and placed it so that when it burned low enough, it would ignite the paper and mattress. According to the prosecution, Sharpe then lit the candle, and both he and Strickler exited the hotel and headed over the mountain. The building would be engulfed in flames only after they had made their escape.

Shortly after the fire, the district attorney said, Wyatt visited Albany and was seen by an impartial witness to pay Sharpe cash. Later, Wyatt would pay other sums, amounting in total to $500. Why would Wyatt be giving these large sums to a former employee? The answer was clear: for a job well done. According to the prosecution, Wyatt's motive for the crime was to collect insurance money on the antiquated hotel, which was having major sewage problems and was being investigated by the state health department.

District Attorney Bascom finished on an emotional note. The *Post Star* described his conclusion this way: "District Attorney Wyman S. Bascom summing up was considered by lawyers as remarkable. His address to the jury will long live in the memory of the large crowd that listened to his unbounded power in oratory."

He reminded the jury that Sharpe, his traveling companion Wallace Strickler and Cornelia Gries had all sworn under oath that this is what transpired. Numerous independent witnesses corroborated parts of their testimonies. There could be no doubt that all the evidence indicated that William Wyatt was guilty.

The lawyers finished their summations at noon, and Judge Rogers declared a recess until 1:00 p.m. The jury was placed in the hands of a court officer and allowed to go to lunch.

SATURDAY AFTERNOON, MAY 19, 1917

The court resumed its session shortly after 1:00 p.m. with Judge Rogers delivering the charge to the jury. One account referred to his instructions as "splendid and fair"; another account called his words to the jury "the most fair and impartial charges ever delivered to a jury."

He started by complimenting the twelve men for their concentration throughout the proceedings and for giving equal consideration to the defense and the state. He cautioned the twelve to make their decision soberly and with deliberation.

Rogers began by reviewing the deliberations leading up to the judicial indictment secured against Wyatt and Sharpe. He stated that according to the evidence offered at the trial, Wyatt could have been indicted for arson in the first degree and could have been tried for that crime because it was shown that human beings were in the building when it was burned down. He dwelt on the manner of an accomplice or accomplices and instructed the jury that the evidence of such a person or persons must be corroborated by other testimony. He said, "Strickler's story, if you believe it, would make him an independent person, but if he saw Sharpe fire the hotel, he was a party to the deed." Likewise, the testimony of Mrs. Cornelia Gries "was corroborative unless she was a party to the crime."

In regards to Bob Sharpe, the judge declared that he "was not an accomplice."

The judge continued his instructions to the jury:

> *The defendant swore that he was in Saratoga at the time Sharpe testified he was in Albany. Mr. Wyatt swore that he never told Sharpe to burn the hotel, that he never gave him any money. The defendant told you he suffered a large and substantial loss. It is not necessary for you gentlemen to find a motive for the alleged crime. Witnesses of high standing offered evidence of the good character of the defendant which raises a presumption of innocence. Mr. Buckell, the former owner of the hotel, was not an interested witness and the same can be said of Mr. Sherman. It would have been necessary for Mrs. Gries to have aided and abetted the crime to become an accomplice.*

Rogers concluded his statement to the jury with this final admonition: "It is for you gentlemen to determine whether the tale of conversations are sufficient to convict the defendant with these crimes."

The judge's remarks lasted approximately forty minutes, and according to handwritten notes of Wyman Bascom, the jury retired from the courtroom and began deliberations at exactly 2:40 p.m.

The jury room was located directly behind the jury box, on the left as one faced the front of the courtroom. The doorway leading into it had a rounded arch above the door, and the room itself had a high ceiling and two windows overlooking the back of the building. There was a large table in the room for the jurors to confer at.

The reporters covering the trial thought that the evidence presented by both sides would cause the deliberations to take some time and that, at its conclusion, the case appeared to be very close. Thus, the speed with which the jury returned took everyone by surprise.

Bascom's notes state that the "jury returned at 4:30 p.m., came with/for instructions." After their short conference with Judge Rogers, they returned to the jury room for more discussion.

Finally, after deliberating for a total time of exactly two hours and thirty-two minutes, the jury foreman notified Judge Rogers that the jury had reached a verdict. All parties were summoned back to the courtroom, and the jury delivered its decision. Bascom's notes were exceedingly precise, stating that the jury came in at 5:12 p.m. Filing back into the jury box, the judgment was read by the foreman at 5:20 p.m.

Surrounded by members of his family, who stood loyally by him during the trial, William H. Wyatt heard the verdict of not guilty ring through the courtroom.

Upon hearing the decision, Wyatt and members of his family were overcome with emotion. They all shed tears of joy while the jury's ruling was applauded by those in the courtroom.

When William H. Wyatt left the courthouse on Saturday, May 19, 1917, he most likely traveled on Route 4 through Fort Edward to his home in Troy. The Esso station on the right is now a Stewart's convenience store. *Courtesy of the Old Fort House Museum/Fort Edward Historical Association.*

The *Glens Falls Times and Messenger* account concludes with the following opinion on the week's events:

> *To those who followed the case in detail it was evident that the prosecution had not established the guilt of the defendant, regardless of the excellent preparation District Attorney Bascom had made, and the splendid manner in which he presented his case. Wyatt was ably represented by Attorney W.H. Williams, who had prepared a strong defense and J.A. Kellogg, who conducted the trial. However at its conclusion, the case appeared to be so close, that a disagreement was expected.*

11

The Pictures

The story now ends where it began, with the pictures found glued to the back of the Lincoln illustration. What is compelling about the nineteen pictures is that they show a period of some months during the winter of 1915–16. They were clearly not all taken on the same day but instead show the building of the new Hulett Hotel over the course of the winter. The first picture shows the aftermath of the fire, and the last picture looks like it was literally taken days before the new hotel opened in the spring.

William Wyatt; his wife, Ella Haynes Wyatt; and his son, Arthur, each appear separately in a photograph. Many workmen and laborers can be seen hard at work. If the same pictures had been left in a time capsule, they would accurately demonstrate the entire process of building the new Hulett Hotel. Was that the intended purpose of whoever put them there? We simply can't say, but we certainly have someone to thank for this historical treasure-trove.

The photographs attached to the back of the Lincoln illustration ultimately leave more questions than answers:

- Who took the pictures and why were they taken?
- Were the pictures intended to be private keepsakes, or were they meant to publically document the building of the new hotel?
- Could these pictures have potentially played a role in the trial?
- Were the pictures' existence even known at the time of the trial?
- Why were the pictures glued to the back of the Lincoln illustration in the first place?

- Does the illustration on the front convey some type of message?
- Where was the illustration originally hung, and would this fact, if known, ultimately point back to the person who attached the pictures to it?
- Photography in the early 1900s was an expensive undertaking. Does the existence of nineteen high-quality images from this time period indicate that the person taking the pictures was a person of wealth?
- Does John D. Sharpe appear in any of the pictures? Sharpe was reported to have run a livery business, and some of the pictures show men on horses pulling lumber. Was Wyatt's chief accuser captured on film?

What makes this story interesting is that we don't know the answers to these questions, but we have the actual pictures to recount vividly a time that would have otherwise been forgotten. The mystery of who glued the pictures to the illustration remains, but the singular act of attaching them to the Lincoln illustration has opened a kaleidoscope into history that would have otherwise been forgotten.

12

Aftermath

John D. Sharpe

The Accuser

Although Wyatt was acquitted, John Sharpe had also been charged with arson in the second degree for, in his words, "firing" the Hulett Hotel. On Monday, May 21, 1917, when stories began appearing in some outlying newspapers regarding the verdict in the Wyatt case, the *Glens Falls Post Star* carried an interesting account of what was in store for Sharpe. "Although [Sharpe] waived immunity, it is generally believed that the District Attorney's office will recommend to the judge that he escape punishment in view of the evidence offered at the trial of Wyatt."

However, this was not the end of the matter for Sharpe. On May 22, 1917, the *Glens Falls Post Star* reported that Attorney Horace E. McKnight of Ballston Spa appeared for Sharpe and that a motion would be made to grant Sharpe his freedom in return for posting $2,000 bail. However, the article stated that this would be based on the agreement of the district attorney and approval of Judge Rogers. The *Post Star* reporter wrote that anything Judge Rogers might do was "conjecture" at that point. The article concluded by stating that "the indictment hanging over Sharpe, will be considered before court adjourns this week."

However, Sharpe was not as fortunate as he would have hoped when the court addressed his indictment at the end of the week. The *Post Star* of Friday,

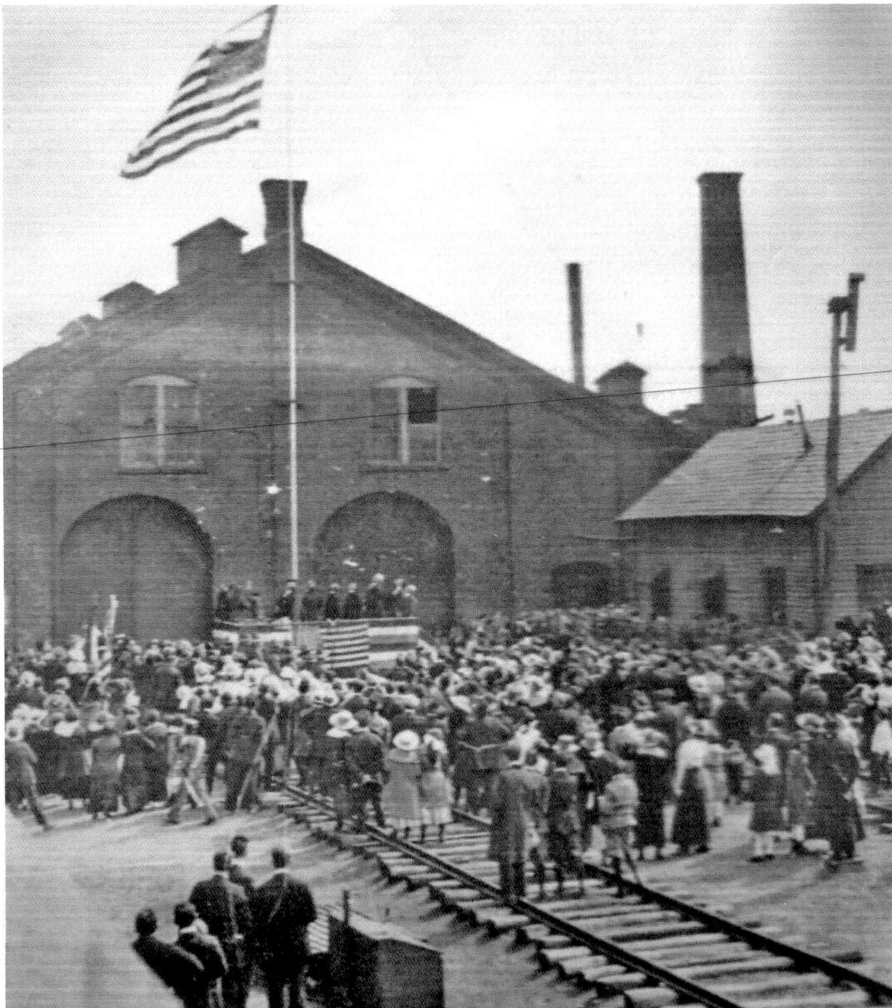

Flag raising ceremony on July 4, 1917, at the D&H shop in Whitehall, New York. Sharpe remained incarcerated in the Washington County jail during the summer of 1917. *Courtesy of the Whitehall Historical Society/Ray Rose.*

May 25, 1917, reported that while Bascom insisted that the sentence, if any, imposed on Sharpe be very light in view of the evidence he had offered at the trial, Judge Rogers did not agree and did not "view the condition in the same light as the county prosecutor." Sharpe was returned to the county jail in Salem "until the fall term of county court at Salem."

So while Wyatt left the Hudson Falls courthouse a free man, the *Post Star* reported on May 28, 1917, that Sharpe was hauled off in shackles to the county jail in Salem to linger for a number of months more.

The last known account of Sharpe's fate comes from a handwritten note found on the back of the original indictment in the Washington County archives:

> *Sept. 10, 1917 in Supreme Court Salem. NY Justice George R. Salisbury presiding, Horace E. McKnight and District Attorney Bascom appear before the Court in the case of* People vs. John D. Sharpe*: Draft—present.*
>
> *Mr. McKnight moves to transfer the Indictment of John D. Sharpe from the County Court to this Court. Motion duly granted.*
>
> *Mr. McKnight moves for dismissal of Indictment. Dist. Atty. consenting thereto.*
>
> *The Court orders Judicial Indictment dismissed and jail dept. discharged and bail ordered discharged. G.W. Curry Clerk*

Thus, Sharpe walked out of the Washington County jail on Monday, September 10, 1917, after spending the summer of 1917 incarcerated. His indictment was dismissed, and he was never tried. He disappears from history at this moment, with no further news accounts referencing his fate.

Cornelia Gries and Wallace Strickler

The Witnesses

Cornelia Gries seems to have returned to her husband and son sometime after the trial. The 1920 census records her as living in Schenectady with the same husband to whom she was married in 1910. At the time of the 1920 census, she was still housing boarders at her residence. The name of her one renter, as of the 1920 census, is recorded as Adolf Ludwig, a thirty-three-year-old collector of insurance who was born in Germany.

Wallace H. Strickler, Sharpe's traveling companion on the night of the fire, is more difficult to track during the years that followed the trial. The 1920 census does not show an exact match for the name Wallace H. Strickler anywhere in the United States. However, an H.W. Strickler is recorded as living in Gloversville, New York. It is possible that the person taking the census simply recorded the first two initials in the wrong order. Gloversville lies west of Albany, and Strickler admitted in 1917 that he worked in the Albany area. Also, the age recorded for the person named H.W. Strickler in the 1920 census coincides with the age Wallace Strickler would have been in 1920.

If this is the same person who testified for the state during the trial, then Strickler, in 1920, was employed as a foreman for an ice company.

These few facts are all that remain about the two principal witnesses who testified for the prosecution. They likewise disappear at this moment from history, with no reference appearing about either in the 1930 U.S. census.

WYMAN S. BASCOM

Washington County District Attorney

Wyman S. Bascom would go on to have a very successful career as Washington County district attorney and ultimately would become Washington County judge. After the tragic death of his wife, Ester, in 1916, he married Julia L. Dobson in 1921.

The 1894 New York State Constitution required that the question "Shall there be a convention to revise the Constitution and amend the same?" be put to voters during 1936. The legislature further clarified this constitutional requirement, and on November 3, 1936, New York voters approved the holding of a constitutional convention. This convention was held in Albany from April 5 to August 26, 1938.

As delegates to the convention, voters chose ninety-two Republicans, seventy-five Democrats and one member of the American Labor Party. Bascom would go on to serve as one of the Republican delegates from the Thirty-third District. Included among the delegates to the convention were Alfred E. Smith, Robert E. Wagner, Hamilton Fish and Robert Moses. Frederick E. Crane, chief judge of the state court of appeals, was chosen as president of the convention.

While failing in the ambitious goal of creating an entirely new constitution acceptable to voters, the 1938 convention did succeed in getting voter approval of a number of significant amendments to the constitution. The most significant accomplishment of the delegates was to add authorization for expanded state government responsibility for social welfare programs.

During the 1930s, both the federal and state government began new programs to assist people caught in the economic depression in the state. The new amendments clarified the state government's role in providing for these programs and helped avoid problems when the state began to take over added responsibility for social programs later on.

Among the many important issues addressed by the convention, two of the most far reaching were: 1) clarifying the legislature's role in taxation and giving it more power in assessing taxes and 2) giving the legislature authorization for transporting children at public expense to and from any school, public or private, in the state.

In 1941, Bascom was again involved in a car accident. The *Adirondack Record/Elizabeth Post* of August 21, 1941, reported that "he was seriously injured" but was "making a steady recovery" and was recuperating at "his camp on Lake George, where he went after leaving Glens Falls Hospital."

In 1944, Bascom was mentioned as a congressional candidate for the new Thirty-sixth Congressional District. However, once he was publically identified as a candidate, he stated that he "was not a candidate and never [was one]." He retired as Washington County judge in 1953 and announced that he would devote himself to the private practice of law.

Bascom died in 1966 at the age of eighty-one. Sadly, almost all of his private papers were disposed of after his widow's death. He is buried in Union Cemetery in Fort Edward, New York. His son, Robert Williams Cowles Bascom, who was six years old on the fateful day his mother died, followed in his father's footsteps and became a New York State Supreme Court justice under Governor Nelson Rockefeller.

Joseph A. Kellogg

Defense Counsel for Wyatt

Joseph A. Kellogg would go on to serve as chairman of the New York State Democratic Party from 1918 to 1919 and became counsel to New York governor Alfred E. Smith in 1919. He would also serve as a member of the platform committee at the Democratic National Convention in 1924, which was the longest continuously running convention in United States political history.

The 1924 Democratic National Convention, held at Madison Square Garden in New York City from June 24 to July 9, took a record 103 ballots

Frank B. Kellogg, a distant relative of Joseph A. Kellogg, was President Coolidge's secretary of state. He co-authored the Kellogg-Briand Pact, for which he was awarded the 1928 Nobel Peace Prize. *U.S. Library of Congress.*

to nominate a presidential candidate. It was also the first national convention in which a major party had a woman, Lena Springs, placed in nomination for the office of vice president. Initial outsider John W. Davis eventually won his party's nomination, as a compromise, after a virtual war of attrition between front-runners William Gibbs McAdoo and Alfred E. Smith. Kellogg, as a delegate from New York, would have been a backer of Smith. Davis went on to be defeated by incumbent president Calvin Coolidge in the presidential election of 1924.

In an ironic twist, Frank B. Kellogg, one of Joseph A. Kellogg's distant relatives who had moved with his family from New York to Minnesota years before, was selected by President Calvin Coolidge to be his secretary of state. He co-authored the Kellogg-Briand Pact, for which he was awarded the Nobel Peace Prize in 1929. Proposed by its other namesake, French foreign minister Aristide Briand, the treaty intended to provide for "the renunciation of war as an instrument of national policy."

Joseph A. Kellogg died at the age of sixty-four on September 8, 1929, in Glens Falls Hospital after undergoing an emergency appendectomy. Dr. C.R. Hoffman was called to the Kellogg home at 6:00 p.m. and found Kellogg seriously ill with appendicitis. He had him taken immediately to the hospital. Dr. Hoffman operated and found that Kellogg's appendix had burst and peritonitis had set in. Kellogg died two hours later, at 8:00 p.m.

Expressions of sympathy instantly poured in from numerous officials and friends. The *Glens Falls Times* of September 9, 1929, quoted Governor

Franklin Roosevelt and former governor Alfred E. Smith among the many who expressed their condolences. Supreme Court justice Erskine C. Rogers said the following:

> *I have known him intimately for many years. His brilliancy, wit and humor, his opinion of men and affairs made him splendid company, always in demand. I once heard him say that the career of a brilliant trial lawyer was soon forgotten like a comet flashing across the sky and disappearing into darkness. He has left, however, not only his fame as a trial lawyer but his record in the departments of state and as a jurist to be long remembered.*

Erskine C. Rogers

Washington County Judge

The picture of Judge Erskine Rogers that ran with his obituary. *Crandall Public Library.*

Judge Rogers went on to be the Republican candidate for New York attorney general in the 1922 election but was defeated when the Alfred E. Smith forces were swept into office, unseating Governor Nathan L. Miller. On November 9, 1922, Governor Miller appointed Judge Rogers as a commissioner to the Hudson River Regulating District, and at the organizational meeting of the board, he was elected chairman.

Judge Rogers was elected to the New York State Supreme Court in 1928 and achieved wide prominence between 1935 and 1938 for his handling of the politically celebrated Druckman murder trial in Brooklyn.

On the night of March 3, 1935, Brooklyn police found the warm, freshly strangled body of Samuel Druckman stuffed in a bloody canvas bag in the trunk of a Ford sedan in the garage of Meyer Luckman, a reputed mobster. Mr. Druckman was Mr. Luckman's brother-in-law and bookkeeper. Caught in the garage, with blood on their hands and clothes, were three men: Meyer Luckman; a nephew, Harry Luckman; and an ex-convict employee named Fred Hull.

When a Kings County grand jury failed to indict any of the three on murder charges, a scandal ensued. New York's governor, Herbert Lehman, ordered a special grand jury investigation. Out of this investigation, a sordid tale of jury tampering, racketeering, political corruption and bribery emerged. Justice Rogers was selected by Governor Lehman to preside over the extraordinary session of Kings County Supreme Court that heard the famous case and ultimately uncovered heinous political corruption and criminal activity.

Justice Rogers's Supreme Court term would have expired on December 31, 1942, but he had a sudden heart attack while at home in Hudson Falls on November 1, 1940. He died two days later, on November 3, 1940, in Glens Falls Hospital.

WILLIAM H. WYATT

After the trial, William H. Wyatt returned to the hotel business he so loved, but not to Huletts Landing. Many newspaper accounts exist of business ventures he would undertake in the few remaining years of his life. Most of these accounts refer to him as a Troy hotel man.

Shortly before his arson trial started in 1917, it was reported in the *Adirondack Record* that Wyatt was converting the Hall House, which for almost half a century had been the leading hotel in Whitehall, into living apartments. The article said the renovation "is planned to have fourteen flats with modern improvements." The article went on to say that "the change leaves only one first rate hotel in Whitehall, the Arlington hotel, opposite the Delaware & Hudson [Railroad] station."

The *Essex County Republican* reported on April 15, 1921, that Wyatt purchased the Hotel Worden and the Arlington Hotel in Lake George from former Lake George supervisor Edwin J. Worden, who was one of the people who posted Wyatt's bail at the time of his arrest. The deal also included a five-year lease of the Worden Annex on the opposite side of the street from the Worden Hotel. The article states that Mr. Worden wanted to devote his time to other interests. These included farming and running several other rooming houses but also improving the water system that supplied the village of Lake George.

Wyatt would not live to see the end of the lease for the Worden Annex. He died at the age of sixty-one on May 31, 1922, in Troy, New York. The official cause of death is recorded in cemetery records as "angina pectoris." This condition, commonly known as angina, is characterized by severe chest pain due to a lack of blood and hence a lack of oxygen to the heart muscle. Generally, it is due to an obstruction or spasm of the coronary arteries. It is usually precipitated by increased physical activity or severe stress. Other risk factors for this condition can include diabetes, family history of coronary heart disease, high blood pressure, high cholesterol, not getting enough exercise, obesity and/or smoking.

Did the stress from the trial kill William Wyatt? Probably not in the short term, but it most likely contributed to the condition that ultimately took his life.

The Wyatt family headstone in Troy's Oakwood Cemetery is not far from the grave of "Uncle Sam" Wilson. *Kapusinski family collection.*

The *Troy Evening Record* of June 1, 1922, carried Wyatt's obituary, which stated that he died suddenly and that a private funeral was to be held in his residence, the Trojan Hotel, at 43 Third Street in Troy. He was buried on June 2, 1922, in Oakwood Cemetery in Troy, New York, one of the largest cemeteries in the United States. Frank P. Hines is listed as the funeral director who handled the interment. Wyatt's wife, Mrs. William H. Wyatt, is listed as his nearest living relative on cemetery records.

The Wyatt family plot is located in a shaded spot with many trees and a large family headstone that simply reads, "Wyatt." It is not far from the grave of "Uncle Sam" Wilson. Buried alongside William H. Wyatt is his wife, Ella Haynes Wyatt, and daughter, Maude Wyatt Granville. Also buried nearby are his son, Arthur Wyatt, and his wife, Mary Wyatt.

THE GLENWOOD HOTEL

After the Glenwood Hotel fire, William Wyatt sold the property to Ellis N. Northrup of Castleton. Northrup was quoted as saying that he wanted to have a hotel built on the Glenwood site and would try to interest someone in the project. However, in 1918 he sold the remaining Glenwood cottage, which originally housed the Glenwood staff, to Jeremiah Durick.

The Glenwood Hotel's last year of operation was 1912. *Courtesy of Donald H. Thompson.*

The site of the Glenwood Hotel as it looks today. *Kapusinski family collection.*

From the time of purchase until his death in 1954, Durick made a hobby out of restoring the cottage. He jacked up the foundation several times, as the cottage began to slide down the hillside.

After Jeremiah Durick passed away in 1954, the former Glenwood cottage passed to his daughter, Catherine Durick Lull. Sadly, it was neglected and was broken into several times during the 1970s. It remained in the Durick/Lull family for over eighty years until it was sold to Mr. and Mrs. Thomas Doran in 2000.

Today, the cottage has been restored, and when a new septic system was installed, the present and past collided. Numerous dishes and bottles were found as the excavation reached what would have been the cellar of the Glenwood Hotel.

Remnants of the Glenwood's tennis court still exist on the hillside behind the property, and a metal pipe that ran to the former washhouse can be seen. The site of the actual hotel is a now a manicured lawn next to the Doran residence.

THE NEW HULETT HOTEL

The hotel that was built to replace the one that burned in 1915 had a long and glorious history. William Wyatt's son, Arthur, went on to operate the Hulett Hotel until his death in 1946. At that time, his estate sold the hotel, grounds and cottages to a group of hotel vacationers who incorporated the operation under the name Huletts Landing Corporation.

After five years of sometimes contentious operation, which was characterized by a good deal of disagreement among the directors about how things should be run, one of the shareholders, George H. Eichler, and his wife, Margaret, purchased the assets of the corporation in January 1952.

Another significant fire would again strike Huletts Landing in the summer of 1953, when the Huletts Casino, which stood directly in front of the hotel, caught fire and burned to the ground. The casino would be rebuilt in another location, but the hotel would finally succumb to age, changes in vacation patterns and tax policy in 1959 after forty-three summers of operation. After getting embroiled in a tax disagreement with the Town of Dresden regarding the hotel's assessment, the Eichlers had the hotel taken down in the winter of 1959–60.

The new Hulett Hotel would go on to be a popular tourist destination for many years. *Kapusinski family collection.*

Tennis courts and a community pavilion now occupy the site of the 1915 fire. *Kapusinski family collection.*

The wood from the hotel went into building many of the homes in Huletts Landing that still stand today. A few private homes occupy the spot where the new Hulett Hotel stood. Tennis courts and a community pavilion, owned by a local homeowners association, are located on the site of the November 1915 fire. The community thrives today with summer vacationers who come to relax on the peaceful and beautiful waters of Lake George.

The Washington County Courthouse

Today, the Washington County courtroom occupies the new courthouse complex in Fort Edward, along with the Washington County Sheriff's Department and the county jail. The new complex was built in 1993. The Hudson Falls courthouse where the trial took place is now privately owned, after having been sold by the county years ago. The first floor houses a restaurant; the actual courtroom is now a performance area for local entertainers.

The 1917 courthouse today houses a restaurant and a performance area for local entertainers. *Kapusinski family collection.*

The courtroom as it looks today. The judge sat in the center underneath the screen, and the jury box was where the piano is now located. *Kapusinski family collection.*

While it is different today in many ways, the basic structure and elegance of the courtroom can still be seen. The high ceilings, the benches where the public sat and the intricate flower molding on the walls are all exactly the same as when the trial took place in 1917. Even the jail cells where the prisoners awaited their trials, while off-limits, are still located in the basement.

William H. Seward

Secretary of State for Presidents Lincoln and Johnson

After surviving the assassination attempt on his life the night President Lincoln was murdered, William Seward made a complete recovery. He would continue to serve as secretary of state under President Andrew Johnson.

Seward's most famous achievement as secretary of state was his successful acquisition of Alaska from Russia. On March 30, 1867, he completed negotiations for the territory, which involved the purchase of 586,412 square miles of territory (more than twice the size of Texas) for $7,200,000, or approximately two cents per acre. The purchase of this frontier land was

Painting by Emanuel Leutze depicting William Seward (seated, left of globe) and Eduard de Stoeckl negotiating the Alaska Purchase. *Alaska State Library.*

alternately mocked by the public as "Seward's Folly," "Seward's Icebox" and Andrew Johnson's "polar bear garden." Today, the state of Alaska celebrates the purchase on Seward's Day, the last Monday of March. When asked what he considered his greatest achievement as secretary of state, Seward replied, "The purchase of Alaska—but it will take the people of the United States a generation before they realize it."

Seward retired as secretary of state after Ulysses S. Grant took office as president. During his last years, he traveled and wrote prolifically. Most notably, he traveled around the world in fourteen months from July 1869 to September 1871. On October 10, 1872, Seward died at the age of seventy-one in his office in his home in Auburn, New York, after having difficulty breathing. His last words were to his children, saying, "Love one another."

Epilogue

Once again, it was not my intent to prove the guilt or innocence of any particular person(s) or to challenge the verdict issued in the arson trial that resulted from the fire. As a person who loves history, my goal was to share the intriguing pictures that were given to me and pass along my research. I've always believed that history is not only imaginatively enriching, but also, and more importantly, it creates meaning as it helps us to understand ourselves in relation to time. I hope you have enjoyed this work, have learned new things, or are perhaps mulling over the evidence presented herein.

However, the story is now over. It is actually long over. The events chronicled in this book at the time it goes to publication will be almost one hundred years old. All the participants in the trial are deceased. In spite of this, people continue to ask me my opinion regarding the trial's verdict.

This is my answer. William H. Wyatt was found not guilty by a jury of his peers. The *Glen Falls Times and Messenger* opined that while "a disagreement was generally looked for," in reality the verdict was returned quickly. The jury spoke loudly. Case closed. The jury members were there; they heard the testimony, they saw the characters and they judged for themselves. At the end of the trial, Wyatt was declared not guilty. When he left the courtroom on the afternoon of Saturday, May 19, 1917, William H. Wyatt, in the eyes of the law, was and forever remains an innocent man.

U.S. Library of Congress.

After examining the case and reviewing the newspaper accounts, it is my opinion—although it was uncorroborated by the testimony offered—that the jury most likely viewed Cornelia Gries as an accomplice to Sharpe. Sharpe testified that Gries had posed as Mrs. Smith on his reconnaissance trip to Huletts Landing before the fire and that she had accompanied him to Keeler's Hotel when the alleged offer was made. Judge Rogers, in his charge to the jury, stated that the testimony of Mrs. Cornelia Gries "was corroborative unless she was a party to the crime."

Could the jury have viewed Wallace Strickler as an accomplice? Judge Rogers's charge to the jury addressed this, too. He said, "Strickler's story, if you believe it, would make him an independent person but if he saw Sharpe fire the hotel he was a party to the deed." Strickler claimed on the witness stand that Sharpe had paid him $100 to accompany him to Huletts, and he said he saw Sharpe light the candle that ultimately burned down the hotel. Even after these admissions, Strickler still claimed that he did not realize that Sharpe was "firing" the hotel.

Did the jury see the actions of Gries and Strickler as those of impartial parties or accomplices to Sharpe out to save themselves? Strickler admitted that he had consulted an attorney after the fire. Gries was seen late at night in the district attorney's office. There is no doubt that, given Judge Rogers's instructions, this was a central issue for the jury to deliberate.

Is this the central issue that swung the pendulum in Wyatt's favor? I simply cannot say.

Wyatt's defense was successful in questioning the characters of all of the accusers while also producing strong witnesses who testified in favor of Wyatt's reputation and integrity. While Wyatt had a former district attorney, Jarvis P. O'Brien, and other local businessmen attest to his character, the defense brought into the trial some interesting information regarding the accusers, especially Cornelia Gries. Wallace Strickler's testimony, when cross-examined by Kellogg, that Gries threatened to "fix the both of you criminals" and that Sharpe told Strickler that Mrs. Gries had "burned a couple of babies or something like that" must have damaged all three accusers' credibility in the minds of the jurors.

While attempting to learn more about Cornelia Gries and this later comment in particular, I did learn that there is an infant female child buried in the Ballston Spa village cemetery also named Cornelia Gries, who died in 1911 at less than one year old. While I could not draw any conclusions from this information, could this child be one of the babies to whom Sharpe's comment alluded?

If Gries's own child was one of the babies Sharpe referred to, then Wyatt's defense might not have wanted to go any further, fearing a sympathetic reaction from the jury to a mother who had lost a child. Likewise, the prosecution might not have delved into this issue because the circumstances surrounding the child may have painted Gries, one of its principal witnesses, in a highly negative way. I could not glean any further information, but this detail certainly raises more questions.

Another central question I was unable to answer definitively was how the investigation leading to the indictments of Wyatt and Sharpe formally began. My own theory is that Strickler's attorney, F.F. Stone, who also worked for one of the insurance companies, may have contacted the authorities, and the investigation led to Cornelia Gries and then to Sharpe. What lends credibility to this theory is that Sharpe was indicted and Strickler wasn't. Could it be that Strickler's information, passed through his attorney, bought him immunity? Additionally, Sharpe testified that F.F. Stone had been present when Sharpe made his statement to the authorities, which in itself seems highly unusual. Another credible possibility is that Gries, who had had a falling out with Sharpe, went to the authorities herself. This is entirely conjecture on my part, though, and the true facts about how the investigation began remain unknown.

What is clear is that Wyatt left the courtroom at the end of the trial a free man, having been found by a jury of his peers not guilty. Sharpe, on the other hand, was hauled off in shackles and left to linger in the county jail

The new Hulett House Hotel as it appeared in 1916. *Courtesy of the Adirondack Museum.*

for a number of months. While Sharpe, too, ultimately walked free, Judge Rogers, by sending him back to jail during the summer of 1917, appears to have wanted to convey some type of message to Sharpe.

The pictures still exist, however, and we have someone unknown to thank for that. Someone took the pictures, and someone attached them to the back of the Lincoln illustration. Perhaps the same person did both, or maybe the pictures were taken by one person and attached by another. There is simply no evidence to let us know for sure. Ironically, that is how history weaves its timeless web. Something known and something unknown are bound together through time. We know something about the pictures, but we don't know the whole story and probably never will. That is the joy of trying to discover history.

So there it is, something as simple as a few pictures glued to the back of an illustration of Abraham Lincoln have taken us back to a raging inferno in a small Adirondack hamlet that was quite similar to one that occurred in a neighboring Vermont town; the building of an exquisite new hotel, which is now also long gone; and a trial that concluded with the words "not guilty" ringing through the courtroom almost a century ago.

Bibliography

Adirondack Record. "Charged with Hiring Man to Burn Lake George Hotel." May 4, 1917.

———. "End of Whitehall Hotel." April 6, 1917.

———. "Twenty in Washington County Jail." March 1917.

Adirondack Record/Elizabeth Post. "Bascom Denies He's Congress Candidate." May 25, 1944.

———. "Erskine C. Rogers Justice of Supreme Court Dies Sunday." November 7, 1940.

———. "Judge Bascom Recovering." August 21, 1941.

———. "W.S. Bascom Resigns as Washington Co. Judge." January 8, 1953.

Buckell, Betty Ahearn. *No Dull Days at Huletts: with Reminiscences of Henry W. Buckell*. Glens Falls, NY: Guy Printing Co., 1985.

Essex County Republican. "Mrs. Wyman S. Bascom, Victim of Auto Accident." May 19, 1916.

———. "Troy Man Buys Lake George Hostelries." April 15, 1921.

Fair Haven Era. "Sudden Death of Hotel Man, William C. Mound." August 31, 1911.

Fort Edward Advertiser. "Salisbury Named." February 15, 1917.

Goodwin, Doris Kearns. *Team of Rivals*. New York: Simon & Schuster, 2005.

Hills, Frederick S. *New York State Men*. Albany, NY: Argus Company, 1910.

History of Washington County, 1878. Philadelphia: Everts & Ensign, 1878.

Glens Falls Post Star. "Hulett House Is Burned to Ground." November 16, 1915.

Glens Falls Times. "Joseph A. Kellogg Dies in Hospital Following Appendicitis Operation." September 9, 1929.

Glens Falls Times and Messenger. "Damaging Evidence Given Against W.H. Wyatt of Troy in Arson Case." May 17, 1917.

———. "Disagreement Is Not Unlikely." May 19, 1917.

———. "John D. Sharpe Testifies that Wyatt Engaged Him to Burn Lake Hotel." May 16, 1917.

———. "Lake George Scene of $50,000 Conflagration." November 15, 1915.

———. "Not Guilty Is Wyatt Verdict." May 21, 1917.

———. "Prosecution Fails to Shake Testimony of Hotel Man Charged with Arson." May 18, 1917.

———. "Wyatt Is Released Under $10,000 Bail." May 1, 1917.

Lake George Mirror. "New Picturesque Huletts." July 1, 1916.

Lord, Thomas Reeves. *Stories of Lake George Fact and Fancy*. Pemberton, NJ: Pinelands Press, 1987.

Moore, Joseph E. *Murder on Maryland's Eastern Shore*. Charleston, SC: The History Press, 2006.

Myers, James Thorn. *History of the City of Watervliet, N.Y., 1630 to 1910*. Troy, NY: Press of H. Stowell & Son, 1912.

New York Times. "Kellogg Is Named Counsel to Smith." December 27, 1918.

———. "Prison Graft Cases Up." October 4, 1913.

Parker, Amasa J., Jr. *The Code of Criminal Procedure of the State of New York*. 17th ed. New York: Banks Law Publishing Company, 1917.

Post Star. "Evidence Complete in Wyatt's Trial." May 19, 1917.

———. "John D. Sharpe in Salem Jail." May 28, 1917.

———. "Montgomery Is Placed on Trial." May 22, 1917.

———. "Much Interest in Sharpe's Trial." May 21, 1917.

———. "'Not Guilty' Is Wyatt's Verdict." May 21, 1917.

———. "Sharpe May Go Back to Salem." May 25, 1917.

Ticonderoga Sentinel. "Complain of Lake George Pollution." February 10, 1916.

———. "Echo of Hulett Landing Fire." May 3, 1917.

———. "G.W. Bush Gets Verdict of $8,000." May 7, 1914.

———. "Kellogg Family One of Oldest in Section." January/March 1925.

———. "Kellogg for Supreme Court." October 5, 1911.

———. "Kellogg Gets His Old Job." January 4, 1912.

———. "Kellogg Nominated by Democrats." September 22, 1904.

———. "Kellogg Will Resign." July 23, 1914.

———. "Killed in Automobile Wreck Near Blue Ridge." May 18, 1916.

———. "Lake George Hotel in Ashes." November 18, 1915.

———. "Lake George Resort Is Sold." January 30, 1913.

———. "Soldiers' Monument Unveiled and Dedicated." July 6, 1916.

———. "Supreme Court Justice Rogers Dead after Short Illness." November 7, 1940.

———. "The Veterans of Post Alfred Weed Announce." April/June 1916.

———. "Whitehall Lawyer Faces Larceny Charge." May 4, 1916.

Thompson, Donald H. *Lake Bomoseen: The Story of Vermont's Largest Little-Known Lake*. Charleston, SC: The History Press, 2009.

Tolles, Bryant F., Jr. *Resort Hotels of the Adirondacks*. Lebanon, NH: University Press of New England, 2003.

Troy Evening Record. "Wyatt." June 1, 1922.

Whitehall Chronicle. "Annual Meeting of Supervisors." November 19, 1915.

———. "Fire Destroys Property." November 19, 1915.

CORRESPONDENCE

Cornelius, James M., curator, Lincoln Collection, Abraham Lincoln Presidential Library and Museum. E-mail, October 22, 2010.

HISTORICAL DOCUMENTS

Adirondack Museum Photographic Collection, Blue Mountain Lake, New York.

Ballston Spa Village Cemetery Records, Ballston Spa, New York.

Crandall Public Library, Glens Falls, New York.

Joseph Doran Collection of Vermont Brochures, Hydeville, Vermont.

Oakwood Cemetery, Historical Records Department, Troy, New York.

Old Fort House Museum/Fort Edward Historical Association, Fort Edward, New York.

U.S. Censuses, 1910, 1920, 1930.

Washington County Archives, Fort Edward, New York.

Whitehall Historical Association, Whitehall, New York.

INTERVIEWS

Doran, Joseph, brother of current owner of the Glenwood property. Friday, August 5, 2011.

Foster, Mike, great-grandson of Willis Foster. Sunday, July 11, 2011.

Huntington, Kathy, Dresden town historian. Sunday, July 11, 2011.

Lowry, Dennis, Washington County archivist. Monday, August 1, 2011.

Manuele, Alex. Thursday, August 5, 2010.

McCarty, Paul, Fort Edward historian. Thursday, December 16, 2010.

Peterson, Agnes, former Dresden town historian and daughter of Royden Barber. Tuesday, June 28, 2011.

Ryan, Bill, Village of Round Lake historian. Friday, December 17, 2010.

Thompson, Donald H. Lake Bomoseen author. Friday, November 12, 2010.

About the Author

George T. Kapusinski is the son of Albert T. and Margaret (Eichler) Kapusinski and the grandson of George and Margaret Eichler, the last owners of the new Hulett Hotel. He enjoys walking, learning about history and writing about local issues on his blog, "The Huletts Current."